WHY MINUS TIMES MINUS IS PLUS

The very basic mathematics of real and complex numbers

Nils K. Oeijord

iUniverse, Inc.

New York   Bloomington

WHY MINUS TIMES MINUS IS PLUS
The very basic mathematics of real and complex numbers

iUniverse books may be ordered through booksellers or by contacting:

iUniverse
1663 Liberty Drive
Bloomington, IN 47403
www.iuniverse.com
1-800-Authors (1-800-288-4677)

ISBN: 978-1-4502-4063-5 (sc)
ISBN: 978-1-4502-4064-2 (ebk)

Printed in the United States of America

iUniverse rev. date: 06/18/2010

PREFACE AND INTRODUCTION

This book is written for a very broad audience. There are no particular prerequisites for reading this book. We hope students of High Schools, Colleges, and Universities, as well as hobby mathematicians, will like and benefit from this book. The book is rigorous and self-contained. All results are proved (or the proofs are optional exercises) and stated as theorems. Important points are covered by examples and optional exercises. Additionally there are also two sections called "More optional exercises (with answers)."

Modern technology uses complex numbers for just about everything. Actually, there is no way one can formulate quantum mechanics without resorting to complex numbers.

Leonard Euler (1707-1786) considered it natural to introduce students to complex numbers much earlier than we do today. Even in his elementary algebra textbook he uses complex numbers throughout the book.

The mathematicians used several thousand years to fully discover the real numbers and the real line. They used several hundred years to fully discover the complex numbers and the complex plane. After hundreds of years of bitter debates the complex numbers finally won acceptance and were put into use in mathematics and physics in Napoleonic times. C. F. Gauss (1777 – 1855) introduced the term 'complex number' in 1831. Raphael Bombelli (1526-1572) was the first to write down how to perform computations with negative numbers, and he was first to give a treatment of any complex numbers. His presentation of the laws of calculation of complex numbers is monumental. He was the one who finally managed to settle the problem with complex numbers. By the way: He introduced $+\mathbf{i}$ and $-\mathbf{i}$, where $\mathbf{i} = \sqrt{-1}$ is the 'imaginary unit'. The lunar crater Bombelli is named after him.

The objective of this text is to give a *clear* exposition of the *real-number system* and the *complex-number system*. A *complex number* consists of the *sum* of a *real number* and an *imaginary number* (see

i

Chapter III). The imaginary number is the *product* of a real number and the *imaginary unit*, written **i**. The imaginary unit and the complex number $z = a + b\mathbf{i}$, where a and b are real numbers, are *calculated* from the *real-number system*, and follows the axioms (and thereby the theorems) of real numbers (as shown in this text). The operations: addition, multiplication, subtraction, and division of complex numbers *follow from* the axioms of real numbers (as shown in this text). Conversely, complex numbers (including the operations) are shown to *satisfy* the axioms (and thereby the theorems) of real numbers. So, even for complex numbers, minus times minus is plus: $(-z)\cdot(-w) = +zw = zw$, and $(-i)(-i) = +ii = i^2$ (see Chapter III). The facts mentioned above prove that *the theorems of complex numbers are basically the same as the corresponding theorems of real numbers.* It's therefore generally *unnecessary* to *directly* prove theorems of complex numbers if we know the corresponding theorems of real numbers. Actually, the theorems (of real numbers) are generally *better* satisfied by complex numbers than by real numbers. Example: In the real-number system the theorem $\sqrt{ab} = \sqrt{a}\sqrt{b}$ is *not* valid if a or b are negative real numbers, but the theorem $\sqrt{zw} = \sqrt{z}\sqrt{w}$, correctly used, is valid for *all* complex numbers z and w. Remember that a real number a is nothing more than the complex number $z = a + 0\mathbf{i}$. Therefore *the theorems of complex numbers are generally valid for real numbers.*

It's a bad habit of many authors to state too many definitions without motivation or proof. This book counters this culture, using the fact that, in mathematics, definitions are often implicitly given by axioms or theorems, or can be calculated directly from axioms or theorems. Instead of bombarding the reader with "manmade" definitions, this book uses, in general, axioms and calculations to do the same job.

The language (symbolism) we use to describe mathematics is not the mathematics itself. The mathematics exists regardless of the symbolism used to describe it. Physical systems can be observed, but mathematical concepts are not observable. Mathematics is all discovered, but the symbolism is invented. Many authors tend to forget that mathematics is *discovered*, and that mathematics is *non-physical* (*non-material*). The Platonic notion is that mathematics is the imperturbable structure that underlies the very architecture of the universe.

Figures (drawings, diagrams, etc) prove nothing and tend to become sleeping pillows working against understanding. Therefore figures are totally lacking in this book. However, drawing useful figures is left as optional exercises.

Note: There are free ***complex number calculators*** on the Internet.
(See also APPENDIX 7.)

Nils K. Oeijord

CONTENTS

Notation

$\{a, b, c, \ldots\}$	the set consisting of the elements $a, b, c, \ldots$.
$x \in X$	x is an element of the set X.
$x \notin X$	x is not an element of the set X.
$X \subset Y$	X is a subset of Y.
N	the set of all natural numbers $(1, 2, 3, \ldots)$.
Z	the set of all integers $(\ldots, -3, -2, -1, 0, 1, 2, 3, \ldots)$.
Q	the set of all rational numbers (*fractions* where numerator and denominator are integers).
R	the set of all real numbers.
C	the set of all complex numbers.
$P \rightarrow Q$	If P then Q (P implies Q).
$Q \rightarrow P$	P if Q.
$P \rightarrow Q$	P only if Q.
$P \leftrightarrow Q$	P if and only if Q (P and Q are both true or both false).
$\forall$	for all; for any; for each.
$a = b$	a equals b.
$a \neq b$	a does not equal b.
$+$	addition sign, plus sign.
$-$	subtraction sign, minus sign.
$\cdot$	multiplication sign.
$\div$	division sign.
$\pm$	plus or minus sign.
$+\infty$	plus infinity.
$-\infty$	minus infinity.
$\dfrac{a}{b}$	a divided by b (the horizontal line is called a *fraction bar*).
$\approx$	is approximate equal to.
$>$	is greater than.
$<$	is less than.
$\geq$	is greater than or equal to.
$\leq$	is less than or equal to.
$\sqrt[n]{a}$ $(a > 0)$	the *positive* nth root of the *positive* real number a, $n = 2, 4, 6, \ldots$.
$\sqrt[n]{a}$ $(a < 0)$	the (negative) nth root of the *negative* real number a, $n = 1, 3, 5, \ldots$.
$\sqrt[n]{z}$ (set)	the n nth roots of the complex number z.
$\sqrt[n]{z}$ (number)	the principal nth root of z.

$\lvert z \rvert$	absolute value of z, the modulus of z.
$\overline{z}$	the complex conjugate of z.
z or w	a typical complex number.
arctan x	inverse of the function tan x.
arg z	argument of z.
Arg z	principal value of arg z.
$\mathbf{i}$	square root of -1.
lim	limit.
log z	complex natural logarithm of z.
Log z	principal value of log z.
ln z	$\log_e z$.
Ln z	principal value of ln z.
$n!$	n factorial; $n! = n \cdot (n-1) \cdot (n-2) \cdot \ldots \cdot 3 \cdot 2 \cdot 1$.
Im (z)	imaginary part of z; Im $(a + b\mathbf{i}) = b$.
Re (z)	real part of z; Re $(a + b\mathbf{i}) = a$.
$\binom{n}{k}$	binomial coefficient, $\binom{n}{k} = \dfrac{n!}{k!(n-k)!}$.
$\left(a_{ij} \right)$	the matrix A.

I   THE ESSENCE OF MATHEMATICS

A Undefined and defined objects

Definitions of concepts are based on undefined concepts. In this book the real numbers are undefined objects, for example. The definitions themselves say nothing as to the existence of the things defined. The existence of each of them must be proved or assured. ***In this book necessary definitions are, in general, taken care of by the axioms.***

B Axioms

The most basic laws of mathematics are the ***axioms***. The axioms are taken for granted. An axiom is independent if it is not a theorem that follows from the other axioms.

C Theorems

From the axioms of mathematics we can prove the ***theorems*** (usual laws) of mathematics.

Once a theorem is proved, it can be used to prove further theorems.

It is crucially important in a proof to use only the axioms and the theorems which have been derived from them, and not depend on any preconceived idea or picture.

Note: It is impossible to *fully* axiomatize mathematics because ***Goedel's theorem*** says that any finite system of axioms is not sufficient to prove every result in mathematics. Goedel proved that mathematics is not ***complete***, that is Goedel proved that ***there exist statements that cannot be proved or disproved***.

D Quotes

"The essence of mathematics is not to make simple things complicated, but to make complicated things simple."
S. Gudder

"The axiomatic method has many advantages over honest work."
Bertrand Russell

"The ultimate goal of mathematics is to eliminate any need for intelligent thought."
Alfred N. Whitehead

"In mathematics you don't understand things. You just get used to them."
Johann von Neumann

"Not to be absolutely certain is, I think, one of the essential things in rationality."
Bertrand Russell

"The problem with simple arguments is that they may be difficult to explain."
Karin Erdman

"As far as the laws of mathematics refer to reality, they are not certain; and as far as they are certain, they do not refer to reality."
Albert Einstein

"Mathematics: The Loss of Certainty."
Title of a book by Morris Kline

"An abstract definition needs at least two interesting examples to justify its existence."
James A. Green

"The unreasonable effectiveness of mathematics in the natural sciences."
Title of an article by Eugene Wigner

"The most incomprehensible thing about the world is that it is at all comprehensible."
Albert Einstein

"I have hardly ever known a mathematician who was able to reason."
Stephen Hawking

"Mathematicians are born, not made."
Henri Poincare

"A lot of mathematicians are a little bit strange in one way or another. It goes with creativity."
Peter L. Duren

"Youngsters from seven countries, asked to come up with a portrait of the typical mathematician, showed a badly dressed, middle-aged nerd with no social life."
The Times, January 3, 2001

"Mathematics is the only science where one never knows what one is talking about nor whether what is said is true."
Bertrand Russell

"Mathematics has the completely false reputation of yielding infallible conclusions. Its infallibility is nothing but identity. Two times two is not four, but it is just two times two, and that is what we call four for short. But four is nothing new at all. And thus it goes on and on in its conclusions, except that in the higher formulas the identity fades out of sight."
Johann Wolfgang von Goethe

"God exists since mathematics is consistent, and the devil exists since we cannot prove the consistency."
Morris Kline

"The physicists defer only to mathematicians, and the mathematicians defer only to God."
L. M. Lederman

"Gentlemen, that is surely true, it is absolutely paradoxical; we cannot understand it, and we don't know what it means. But we have proved it, and therefore we know it must be the truth."
Benjamin Pierce (after proving Euler's identity in a lecture)

"Euler's identity is the most remarkable formula in mathematics."
Richard Feynman

"In the one equation $0 = 1 + e^{i\pi}$ the five fundamental numbers $0, 1, i, e, \pi$ are "interwoven in a truly wonderful manner" which has given rise on occasions to metaphysical speculation."
R. Remmert

"Had $+1$, -1 and $\sqrt{-1}$, instead of being called positive, negative and imaginary […] unity, been given the names, say, of direct, inverse and lateral unity, there would hardly have been any scope for […] obscurity."
C. F. Gauss

"Beauty is the first test; there is no permanent place in the world for ugly mathematics."
G. H. Hardy

"One cannot escape a feeling that these mathematical formulae have an independent existence and intelligence of their own, wiser than we are, … ."
Heinrich Hertz

"Our experience hitherto justifies us in believing that nature is the realization of the simplest conceivable mathematical ideas."
Albert Einstein

"What is it that breathes fire into the equations and makes a universe for them to describe … . Why does the universe go to all the bother of existing?"
Stephen Hawking

"Somehow, there must be, wondered Einstein, a kind of 'pre-established harmony' between human inventive conceptual imagination and aspects of reality itself."
Arne Naess

"Now, my own suspicion is that the universe is not only queerer than we suppose, but queerer than we *can* suppose."
J. B. S. Haldane

"The most beautiful thing we can experience is the mysterious. It is the source of all true art and science."
Albert Einstein

"Equations are more important to me, because politics is for the present, but an equation is something for eternity."
Albert Einstein

"We do not know why they [elementary particles] have the masses they do; we do not know why they transform into another the way they do; we do not know anything!"
George Gamow

"Science is an edged tool, with which men play like children, and cut their own fingers."
Arthur Eddington

"Science without religion is lame, religion without science is blind."
Albert Einstein

II   THE REAL-NUMBER SYSTEM

A The real numbers

In this book the real numbers are taken for granted and not defined. But it is well known that ***the set of all real numbers may be pictured geometrically as a line***, each point of this line representing a real number. The real numbers consist of the positive real numbers, the zero, and the negative real numbers. The ***positive direction*** of this ***real-number line*** is to the right, i.e. in the direction of $+\infty$. Real numbers (and complex numbers) are designated by lower-case letters: $a, b, c, \ldots, x, y, z$.

B Operations on the real numbers

We assume the existence of four operations on the real numbers called addition, subtraction, multiplication, and division. We attach no special meanings to the symbols +, •, ·, and ÷ other than those contained in the axioms.

The reader obviously does have a good understanding and a good working knowledge of addition of real numbers.

Note that ***$a • b = c$ means $b + c = a$, that is $a • b = c$ if and only if $b + c = a$.***

Note that ***$a • b$ means $b + b + \ldots + b$ (a times b).***

Note that ***if $b • 0$ then $a ÷ b = c$ means $b • c = a$, that is if $b • 0$ then $a ÷ b = c$ if and only if $bc = a$.***

We take for granted that for each real number a and b the sum $a + b$ and the product ab do exist, and are unique real numbers, and are well defined by a and b. So we take for granted that the operations of addition and multiplication are ***closed***. (See Axiom 0.) The operations subtraction and division are the undoing of addition and multiplication, respectively, so we make no flat assumptions about subtraction and division. (See Corollaries of Axiom 0.) But note that division of real numbers is *not* closed as we cannot divide by zero. Any real number (except zero) divided by zero is ***undefined***. Zero divided by zero is said to be ***indeterminate***. Zero divided by any real number (except zero) is zero (See Theorem 21.) Zero multiplied by any real number (or any real number multiplied by zero) is zero (See Theorem 19 and Theorem 20.)

C Axioms of the real numbers

Axiom 0 (The laws of existence and uniqueness of sums and products)
***For each real number a and b the sum a + b and the product ab do exist
and is unique.***

Corollaries of Axiom 0: ***For each real number a and b the difference a • b
exists and is unique. For each real number a and each real number b • 0
the quotient a ÷ b exists and is unique.***

Axiom 1 (The law of contradiction)
***Mutually contradictory propositions cannot both be true and cannot both
be false, and no proposition can be at once true and false.***

Axiom 2 (The law of substitution)
If a = b then we can substitute a for b in an equation.

Axiom 3 (The law of symmetry)
(Here an axiom but it is a corollary (immediate consequence) of Axiom 2.)
If a = b then b = a.

Axiom 4 (The law of reflexivity)
(Here an axiom but it is a corollary of Axiom 2. However it's arguably an axiom rather than a
corollary since it has no premises.)
a = a.

Axiom 5 (The law of transitivity)
If a = b and b = c then a = c.

Axiom 6 (The law of the plus sign)
+a = a. (Corollary: ***+0 = 0.***)

Axiom 7 (The law of parentheses)
(a) = a.

Axiom 8 (The commutative law of addition)
a + b = b + a. (Corollary: ***a + b = a + b.***)

Axiom 9 (The associative law of addition)
$a + (b + c) = (a + b) + c.$

Axiom 10 (The commutative law of multiplication)
$ab = ba.$ (Corollary: $ab = ab.$)

Axiom 11 (The associative law of multiplication)
$a(bc) = (ab)c.$

Axiom 12 (The distributive law of addition)
$a(b + c) = ab + ac.$

Axiom 13 (The laws of existence of 0 and 1)
$a + 0 = a$ and $1a = a.$

Axiom 14 (The law of existence of negatives. See also Axiom 15)
$a + (\bullet a) = 0.$

Axiom 15 (The law of existence of additive inverse)
(This is the same axiom as Axiom 14.)
For every real number a there is a real number b such that a + b = 0.

Axiom 16 (The law of existence of multiplicative inverse)
For every real number a $\bullet$ 0 there is a real number b such that ab = 1.

D The real-number system

The real-number system is the totality of the real numbers, the operations on the real numbers, and the axioms of the real numbers. *The real-number system is not mathematically complete.* Example: In the real-number system we cannot solve the equation $x^2 + 1 = 0$. But *the complex-number system includes the real-number system and is mathematically complete.* (See Chapter III.)

E Basic theorems of elementary mathematics

Note: In this book when a theorem is proved it is not 'the' proof. *No attempt is made to search for the most elegant proof.*

Theorem 1 (The law of cancellation for addition)

$a + b = a + c$ if and only if $b = c$.

Proof:

Given $a + b = a + c$
$d + a = 0$ (Axiom 15)
$d + (a + b) = d + (a + b)$ (Corollary of Axiom 8)
$d + (a + b) = d + (a + c)$ (Axiom 2)
$(d + a) + b = (d + a) + c$ (Axiom 9)
$0 + b = 0 + c$ (Axiom 2)
$b + 0 = c + 0$ (Axiom 8, Axiom 2)
$b = c$ (Axiom 13).

Given $b = c$
$a + b = a + b$ (Corollary of Axiom 8)
$a + b = a + c$ (Axiom 2).

Theorem 2
The real number zero is unique.

Proof:

$a + 0 = a$ (Axiom 13)
$a + 0' = a$ (Axiom 13)
$a = a + 0$ (Axiom 3)
$a + 0' = a + 0$ (Axiom 5)
$0' = 0$ (Theorem 1).

Theorem 3
• 0 = 0.

Proof:

$0 + (-0) = 0$ (Axiom 14)
$(-0) + 0 = 0$ (Axiom 8, Axiom 2)
$(-0) + 0 = (-0)$ (Axiom 13)
$(-0) = (-0) + 0$ (Axiom 3)
$(-0) = 0$ (Axiom 5)
$-0 = 0$ (Axiom 7, Axiom 2).

Theorem 4
$+(+a) = a.$

Proof:

$+(+a) = (+a)$ (Axiom 6, Axiom 7)
$+(+a) = +a$ (Axiom 7)
$+(+a) = a$ (Axiom 6).

Theorem 5
$+(\bullet a) = \bullet a.$

Proof:

$-a = -a$ (Axiom 14 ($-a$ exists and is a real number), Axiom 4)
$+(-a) = (-a)$ (Axiom 6, Axiom 7)
$(-a) = -a$ (Axiom 7)
$+(-a) = -a$ (Axiom 5).

Theorem 6
$-(+a) = -a.$

Proof:

$-(+a) = -(+a)$ (Axiom 4)
$-(+a) = -a$ (Axiom 6, Axiom 7).

Corollary of Theorem 6: **$-(ab) = -ab.$**

Theorem 7
−(−*a*) = *a*.

Proof:

$(-a) + [-(-a)] = 0$ (Axiom 14)
$a + (-a) = 0$ (Axiom 14)
$0 = (-a) + a$ (Axiom 3, Axiom 8)
$(-a) + [-(-a)] = (-a) + a$ (Axiom 5)
$[-(-a)] = a$ (Theorem 1)
$-(-a) = a$ (Axiom 7).

Theorem 8
+(−*a*) = −(+*a*).

Proof:

$+(-a) = -a$ (Theorem 5)
$-(+a) = -a$ (Theorem 6)
$-a = -(+a)$ (Axiom 3)
$+(-a) = -(+a)$ (Axiom 5).

Theorem 9
+(+*a*) = −(−*a*).

Proof:

$+(+a) = a$ (Theorem 4)
$-(-a) = a$ (Theorem 7)
$a = -(-a)$ (Axiom 3)
$+(+a) = -(-a)$ (Axiom 5).

Theorem 10
$xa + ya = (x + y)a.$

Proof:

xa + ya = ax + ay (Axiom 10, Axiom 2)
$ax + ay = a(x + y)$ (Axiom 12, Axiom 3)
$ax + ay = (x + y)a$ (Axiom 10, Axiom 2)
$xa + ya = (x + y)a$ (Axiom 5).

Theorem 11
Given a and b, then there is one and only one x such that a + x = b.

Proof:

$a + y = 0$ (Axiom 15)
$x' = y + b$ (Axiom 0)
$a + x' = a + (y + b)$ (Corollary of Axiom 8, Axiom 2)
$a + (y + b) = (a + y) + b$ (Axiom 9)
$a + (y + b) = 0 + b$ (Axiom 2)
$0 + b = b + 0$ (Axiom 8)
$b + 0 = b$ (Axiom 13)
$a + x' = b$ (Axiom 5).

$a + x = b$ (given)
$b = a + x$ (Axiom 3)
$a + x' = a + x$ (Axiom 5)
$x' = x$ (Theorem 1).

Corollary of Theorem 11
Note that $x = b - a$ because $a + x = b.$
Therefore **subtraction is always possible and unique.**
(This is the same law as the Corollary of Axiom 0.)

Theorem 12
Every real number has a unique negative.

Proof:

$a + (-a) = 0$ (Axiom 14)
$a + x = 0$ (Axiom 15)
$0 = a + x$ (Axiom 3)
$a + (-a) = a + x$ (Axiom 5)
$(-a) = x$ (Theorem 1)
$x = -a$ (Axiom 3, Axiom 7)
$a + x' = 0$ (Axiom 15)
$0 = a + x'$ (Axiom 3)
$a + x = a + x'$ (Axiom 5)
$x = x'$ (Theorem 1).

Theorem 13
$a + (\bullet b) = a \bullet b.$

Proof:

$a + (-b) = x$ (Axiom 0)
$[a + (-b)] + b = x + b$ (Axiom 7, Corollary of Axiom 8)
$a + [(-b) + b] = x + b$ (Axiom 9)
$a + [b + (-b)] = x + b$ (Axiom 8, Axiom 2)
$a + 0 = x + b$ (Axiom 14, Axiom 2)
$a = x + b$ (Axiom 13, Axiom 2)
$x = a - b$ (Corollary of Theorem 11)
$a + (-b) = a - b$ (Axiom 5).

Theorem 14
$a + x = b$ *if and only if* $x = b \bullet a.$

Proof:

$a + x = b$ (given)
$(a + x) + (-a) = b + (-a)$ (Axiom 7, Corollary of Axiom 8)
$(a + x) + (-a) = a + [x + (-a)]$ (Axiom 9)
$a + [x + (-a)] = a + [(-a) + x]$ (Axiom 8, Axiom 2)

12

$a + [(-a) + x] = [a + (-a)] + x$ (Axiom 9)
$[a + (-a)] + x = 0 + x$ (Axiom 14, Axiom 2)
$0 + x = x + 0$ (Axiom 8)
$x + 0 = x$ (Axiom13)
$(a + x) + (-a) = x$ (Axiom 5)
$x = (a + x) + (-a)$ (Axiom 3)
$x = b + (-a)$ (Axiom 5)
$x = b - a$ (Theorem 13, Axiom 2).

$x = b - a$ (given)
$x = b + (-a)$ (Theorem 13, Axiom 2)
$a + x = a + [b + (-a)]$ (Corollary of Axiom 8, Axiom 7, Axiom 2)
$a + [b + (-a)] = a + [(-a) + b]$ (Axiom 8, Axiom 2)
$a + [(-a) + b] = [a + (-a)] + b$ (Axiom 9)
$[a + (-a)] + b = 0 + b$ (Axiom 14, Axiom 2)
$0 + b = b + 0$ (Axiom 8)
$b + 0 = b$ (Axiom 13)
$a + x = b$ (Axiom 5).

Theorem 15
$a \cdot 0 = a.$

Proof:

$a - 0 = a + (-0)$ (Theorem 13, Axiom 3)
$a + (-0) = a + 0$ (Theorem 3, Axiom 7, Axiom 2)
$a + 0 = a$ (Axiom 13)
$a - 0 = a$ (Axiom 5).

Theorem 16
$a \cdot a = 0.$

Proof:

$a - a = a + (-a)$ (Theorem 13, Axiom 3)
$a + (-a) = 0$ (Axiom 14)
$a - a = 0$ (Axiom 5).

Theorem 17
0 + *a* = *a*.

Proof:

$0 + a = a + 0$ (Axiom 8)
$a + 0 = a$ (Axiom 13)
$0 + a = a$ (Axiom 5).

Theorem 18
0 • *a* = •*a*.

Proof:

$0 - a = 0 + (-a)$ (Theorem 13, Axiom 3)
$0 + (-a) = (-a) + 0$ (Axiom 8)
$(-a) + 0 = -a$ (Axiom 13, Axiom 7, Axiom 2)
$0 - a = -a$ (Axiom 5).

Theorem 19
$a \cdot 0 = 0$.

Proof:

$a \cdot 0 = a(0 + 0)$ (Theorem 17, Axiom 7, Axiom 2)
$a(0 + 0) = a \cdot 0 + a \cdot 0$ (Axiom 12)
$a \cdot 0 = a \cdot 0 + a \cdot 0$ (Axiom 5)
$a \cdot 0 + [-(a \cdot 0)] = (a \cdot 0 + a \cdot 0) + [-(a \cdot 0)]$ (Axiom 0, Axiom 7, Axiom 3, Axiom 2, Theorem 1)
$0 = a \cdot 0 + \{a \cdot 0 + [-(a \cdot 0)]\}$ (Axiom 14, Axiom 9, Axiom 3, Axiom 2)
$0 = a \cdot 0 + 0$ (Axiom 2, Axiom 14)
$0 = a \cdot 0$ (Axiom 2, Axiom 13)
$a \cdot 0 = 0$ (Axiom 3).

Theorem 20
$0 \cdot a = 0$.

Proof:

$0 \cdot a = a \cdot 0$ (Axiom 10)
$a \cdot 0 = 0$ (Theorem 19)
$0 \cdot a = 0$ (Axiom 5).

Theorem 21
$0 : a = 0$; $a \neq 0$.

Proof:

$a \cdot 0 = 0$ (Theorem 19)
$a \neq 0$.

Theorem 22
$a : a = 1$; $a \neq 0$.

Proof:

$1a = a$ (Axiom 13)
$a \cdot 1 = a$ (Axiom 10, Axiom 2).

Theorem 23
$(+a)(+b) = ab$.

Proof:

$(+a)(+b) = (a)(b)$ (Axiom 6, Axiom 2)
$(a)(b) = ab$ (Axiom 7, Axiom 2)
$(+a)(+b) = ab$ (Axiom 5).

Corollaries of Theorem 23: **$ab = (ab) = +ab = +(ab) = (+ab) = a(+b) = (+a)b$.**

Theorem 24
$a(-b) = -(ab)$.

Proof:

$0 = a \cdot 0$ (Theorem 20)
$a \cdot 0 = a[b + (-b)]$ (Axiom 14, Axiom 2)
$a[b + (-b)] = ab + a(-b)$ (Axiom 12)
$0 = ab + a(-b)$ (Axiom 5)
$ab + a(-b) = 0$ (Axiom 3)
$ab + [-(ab)] = 0$ (Axiom 14)
$0 = ab + [-(ab)]$ (Axiom 3)
$ab + a(-b) = ab + [-(ab)]$ (Axiom 5)
$a(-b) = -(ab)$ (Theorem 1).

Corollary of Theorem 24: **$a(-b) = -ab$.**

Theorem 25
$(-a)b = -(ab)$.

Proof:

$(-a)b = b(-a)$ (Axiom 10)
$b(-a) = -(ba)$ (Theorem 24)
$-(ba) = -(ab)$ (Axiom 10, Axiom 2)
$(-a)b = -(ba)$ (Axiom 5).

Corollary of Theorem 25: **$(-a)b = -ab$.**

Corollary of Theorem 24 and 25: **$(-a)b = a(-b)$.**

Theorem 26 (*Why minus times minus is plus*)
$(-a)(-b) = ab$.

Proof:

$(-a)(-b) = -[a(-b)]$ (Theorem 25)
$-[a(-b)] = -[(-b)a]$ (Axiom 10, Axiom 2)

16

$-[(-b)a] = -[-(ba)]$ (Axiom 25, Axiom 2)
$-[-(ba)] = ba$ (Theorem 7, Axiom 7)
$ba = ab$ (Axiom 10)
$(-a)(-b) = ab$ (Axiom 5).

Theorem 27 (The law of cancellation for multiplication)
If ab = ac and a • 0 then b = c.

Proof:

$ab = ac$ and $a \neq 0$ (given)
$ab + (-ac) = ac + (-ac)$ (Corollary of Axiom 8, Axiom 2)
$ac + (-ac) = 0$ (Axiom 14)
$ab + (-ac) = 0$ (Axiom 5)
$ab + a(-c) = 0$ (Theorem 24, Axiom 2)
$a[b + (-c)] = 0$ (Axiom 12, Axiom 2)
$b + (-c) = 0 / a$ (Corollary of Axiom 0)
$a \cdot 0 = 0$ (Theorem 19)
$0 = 0 / a$ (Corollary of Axiom 0)
$0 / a = 0$ (Axiom 3)
$b + (-c) = 0$ (Axiom 5)
$b + (-b) = 0$ (Axiom 14)
$0 = b + (-b)$ (Axiom 3)
$b + (-c) = b + (-b)$ (Axiom 5)
$-c = -b$ (Theorem 1)
$(-1)(-c) = (-1)(-c)$ (Corollary of Axiom 10)
$(-1)(-c) = (-1)(-b)$ (Axiom 2)
$1c = 1b$ (Theorem 26, Axiom 2)
$c = b$ (Axiom 13, Axiom 2)
$b = c$ (Axiom 3).

Theorem 28
The number 1 is unique.

Proof:

$1 \cdot a = a$ (Axiom 13)
$1' \cdot a = a$ (Axiom 13)
$a = 1 \cdot a$ (Axiom 3)

$1' \cdot a = 1 \cdot a$ (Axiom 5)
$a \cdot 1' = a \cdot 1$ (Axiom 10, Axiom 2)
$a \neq 0$ (given)
$1' = 1$ (Theorem 27).

Note: Formally (because $0/0$ is not a real number):
$a = 0$ (given)
$0 \cdot 1' = 0 \rightarrow 1' = 0/0$
$0 \cdot 1 = 0 \rightarrow 1 = 0/0 \rightarrow 0/0 = 1$
$\rightarrow 1' = 1$.

Theorem 29
Given a and b, and a • 0, then there is one and only one x such that ax = b.

Proof for existence:

$ax = b$, where $a \neq 0$ (given)
$ax' = 1$ (Axiom 16)
$(ax')b = 1b$ (Corollary of Axiom 10, Axiom 2)
$a(x'b) = b$ (Axiom 11, Axiom 13, Axiom 2)
$b = a(x'b)$ (Axiom 3)
$ax = a(x'b)$ (Axiom 5)
$x = x'b$ (Theorem 27).

Proof for uniqueness:

$ax = b$ (given)
$ax^* = b$ (given)
$b = ax$ (Axiom 3)
$ax^* = ax$ (Axiom 5)
$x^* = x$ (Theorem 27).

Theorem 30
If ab = 0 then a = 0 and/or b = 0.

Proof:

(1) Suppose that $ab = 0$, $a \neq 0$, $b = 0$.
$ab = 0$ (given)

$a \cdot 0 = 0$ (Theorem 19)
$0 = a \cdot 0$ (Axiom 3)
$ab = a \cdot 0$ (Axiom 5)
$b = 0$ (Theorem 27)
$a \neq 0$ (given).

(2) Suppose that $ab = 0$, $a = 0$, $b \neq 0$.
$ab = 0$ (given)
$ba = 0$ (Axiom 10, Axiom 2)
$b \cdot 0 = 0$ (Theorem 19)
$0 = b \cdot 0$ (Axiom 3)
$ba = b \cdot 0$ (Axiom 5)
$a = 0$ (Theorem 27)
$b \neq 0$ (given).

(3) Suppose that $ab = 0$, $a = 0$, $b = 0$.
$ab = 0$ (given)
$0 = 0 \cdot 0$ (Theorem 19)
$ab = 0 \cdot 0$ (Axiom 5).

(4) Suppose $ab = 0$, $a \neq 0$, $b \neq 0$.
Proof by contradiction:
$ab = 0$ (given)
$a \cdot 0 = 0$ (Theorem 19)
$0 = a \cdot 0$ (Axiom 3)
$ab = a \cdot 0$ (Axiom 5)
$b = 0$ (Theorem 27)
$b \neq 0$ (given).

Theorem 31
•a = •a.

Proof:

$-a + 0 = -a$ (Axiom 13)
$-a = -a + 0$ (Axiom 3)
$-a + 0 = -a + 0$ (Axiom 5)
$0 + (-a) = 0 + (-a)$ (Axiom 7, Axiom 8, Axiom 2)
$-a = -a$ (Axiom 7, Theorem 1).

Theorem 32
•a = •a if and only if a = a.

Proof:

$-a = -a$ (given)
$(-1)(-a) = (-1)(-a)$ (Corollary of Axiom 10)
$1a = 1a$ (Axiom 26)
$a = a$ (Axiom 13).

$a = a$ (given)
$(-1)a = (-1)a$ (Corollary of Axiom 10)
$-(1a) = -(1a)$ (Theorem 25, Axiom 2)
$-(a) = -(a)$ (Axiom 13, Axiom 2)
$-a = -a$ (Axiom 7, Axiom 2).

Theorem 33
a = •b if and only if b = •a.

Proof:

$a = -b$ (given)
$-b = a$ (Axiom 3)
$(-1)(-b) = (-1)a$ (Corollary of Axiom 10, Axiom 2)
$1b = -1a$ (Theorem 26, Corollary of Theorem 25, Axiom 2)
$b = -a$ (Axiom 13, Axiom 2).

$b = -a$ (given)
$-a = b$ (Axiom 3)
$(-1)(-a) = (-1)b$ (Corollary of Axiom 10, Axiom 2)
$1a = -1b$ (Theorem 26, Corollary of Theorem 25, Axiom 2)
$a = -b$ (Axiom 13, Axiom 2).

Theorem 34
a + (+b) = a + b.

Proof:

$a + (+b) = a + (b)$ (Axiom 6, Axiom 2)

$a + (b) = a + b$ (Axiom 7, Axiom 2)
$a + (+b) = a + b$ (Axiom 5).

Theorem 35
$a \cdot (+b) = a \cdot b.$

Proof:

$a - (+b) = a - (b)$ (Axiom 6, Axiom 2)
$a - (b) = a - b$ (Axiom 7, Axiom 2)
$a - (+b) = a - b$ (Axiom 5).

Corollary of Theorem 13 and 35: $a + (-b) = a - (+b).$

Theorem 36
$a \cdot (-b) = a + b.$

Proof:

$a - (-b) = a + [-(-b)]$ (Theorem 13, Axiom 3)
$a + [-(-b)] = a + (b)$ (Theorem 7, Axiom 2)
$a + (b) = a + b$ (Axiom 7, Axiom 2)
$a - (-b) = a + b$ (Axiom 5).

Theorem 37
$a(b - c) = ab \cdot ac.$

Proof:

$a(b - c) = a[b + (-c)]$ (Theorem 13, Axiom 2)
$a[b + (-c)] = ba + a(-c)$ (Axiom 12)
$ab + a(-c) = ab + [-(ac)]$ (Theorem 24, Axiom 7, Axiom 2)
$ab + [-(ac)] = ab - ac$ (Theorem 13, Axiom 7)
$a(b - c) = ab - ac$ (Axiom 5).

Theorem 38
$a + b = -(-a - b)$.

Proof:

$a + b = 1a + 1b$ (Axiom 13, Axiom 2)
$1a + 1b = (-1)(-a) + (-1)(-b)$ (Theorem 26)
$(-1)(-a) + (-1)(-b) = (-1)[(-a) + (-b)]$ (Axiom 12)
$(-1)[(-a) + (-b)] = (-1)(-a - b)$ (Theorem 13, Axiom 7, Axiom 2)
$(-1)(-a - b) = -[1(-a - b)]$ (Theorem 25)
$-[1(-a - b)] = -(-a - b)$ (Axiom 13, Axiom 7)
$a + b = -(-a - b)$ (Axiom 5).

Theorem 39
$a \cdot b = -(-a + b)$.

Proof:

$a - b = a + (-b)$ (Theorem 13)
$a + (-b) = -[-a - (-b)]$ (Theorem 38)
$-[-a - (-b)] = -(-a + b)$ (Theorem 36)
$a - b = -(-a + b)$ (Axiom 5).

Theorem 40
$\bullet a + b = -(a - b)$.

Proof:

$-a + b = -[-(-a) - b]$ (Theorem 38)
$-[-(-a) - b] = -(a - b)$ (Theorem 7)
$-a + b = -(a - b)$ (Axiom 5).

Corollary of Axiom 8 and Theorem 13: **$b \cdot a = \bullet(a \cdot b)$.**

Theorem 41
$\bullet a \cdot b = -(a + b)$.

Proof:

$-a - b = -[-(-a) + b]$ (Theorem 39)
$-[-(-a) + b] = -(a + b)$ (Theorem 7)
$-a - b = -(a + b)$ (Axiom 5).

The logic of the axioms says:

$x \pm y \pm z = (x \pm y) \pm z$
$x \pm y \pm z \pm u = [(x \pm y) \pm z] \pm u$
$x \pm y \pm z \pm u \pm v = \{[(x \pm y) \pm z] \pm u\} \pm v$
etc.

Theorem 42
$a + (b + c) = a + b + c.$

Proof: Optional exercise.

Theorem 43
$a + (b - c) = a + b \bullet c.$

Proof: Optional exercise.

Theorem 44
$a + (-b + c) = a \bullet b + c.$

Proof: Optional exercise.

Theorem 45
$a + (-b - c) = a \bullet b \bullet c.$

Proof: Optional exercise.

Theorem 46
$$a \cdot (b + c) = a \cdot b \cdot c.$$

Proof. Optional exercise.

Theorem 47
$$a \cdot (b - c) = a \cdot b + c.$$

Proof: Optional exercise.

Theorem 48
$$a \cdot (-b + c) = a + b \cdot c.$$

Proof: Optional exercise.

Theorem 49
$$a \cdot (-b - c) = a + b + c.$$

Proof: Optional exercise.

Theorems 42 – 49 are special cases of the following two theorems:
$$x + (\pm y \pm z \pm u \pm v \pm \ldots) = x \pm y \pm z \pm u \pm v \pm \ldots$$
$$x - (\pm y \pm z \pm u \pm v \pm \ldots) = x \mp y \mp z \mp u \mp v \mp \ldots.$$

Theorem 50 (The general commutative law of addition)
The order in which you add two or more real numbers does not affect the result.

Theorem 51 (The general associative law of addition)
**When you add any three or more real numbers, the grouping of the numbers does
not affect the result.**

The logic of the axioms says:

$$x \cdot y \cdot z = (x \cdot y) \cdot z$$
$$x \cdot y \cdot z \cdot u = [(x \cdot y) \cdot z] \cdot u$$
$$x \cdot y \cdot z \cdot u \cdot v = \{[(x \cdot y) \cdot z] \cdot u\} \cdot v.$$

Theorem 52 (The general commutative law of multiplication)
The order in which you multiply two or more real numbers does not affect the result.

Theorem 53 (The general associative law of multiplication)
When you multiply any three or more real numbers, the grouping of the numbers
does not affect the result.

Theorem 54 (A generalization of Axiom 12)
$$k(a \pm b \pm c \pm \ldots) = ka \pm kb \pm kc \pm \ldots .$$

Theorem 55 (The square of a sum rule)
$$(a + b)^2 = (a + b)(a + b) = a^2 + 2ab + b^2.$$

Proof: Optional exercise.

Theorem 56 (The square of a difference rule)
$$(a \cdot b)^2 = (a \cdot b)(a \cdot b) = a^2 - 2ab + b^2.$$

Proof: Optional exercise.

Theorem 57 (The conjugate rule)
$$(a + b)(a \cdot b) = a^2 - b^2.$$

Proof. Optional exercise.

Theorem 58

$$\frac{a}{1} = a.$$

Proof: Optional exercise.

Theorem 59

$$\frac{a}{a} = 1, \ a \neq 0.$$

Proof: See Theorem 22.

Theorem 60

$$\frac{0}{a} = 0, \ a \neq 0.$$

Proof: See Theorem 20.

Theorem 61

$$\frac{c}{a} \cdot a = c, \ a \neq 0.$$

Proof. Optional exercise.

Theorem 62

$$\frac{a}{b} = c \leftrightarrow \frac{a}{c} = b, \ b \neq 0, c \neq 0.$$

Proof: Optional exercise.

Theorem 63 (*Cross multiplication*)

$$\frac{a}{b} = \frac{c}{d} \leftrightarrow ad = bc, \ b \neq 0, d \neq 0.$$

Proof: Optional exercise.

Theorem 64

$$\frac{ac}{bc} = \frac{a}{b}, \ b \neq 0, c \neq 0.$$

Proof: Optional exercise.

Theorem 65

$$\frac{+a}{+b} = +\frac{a}{b}, \ b \neq 0.$$

Proof: Optional exercise.

Theorem 66

$$\frac{-a}{b} = -\frac{a}{b}, \ b \neq 0.$$

Proof: Optional exercise.

Theorem 67

$$\frac{a}{-b} = -\frac{a}{b}, \ b \neq 0.$$

Proof: Optional exercise.

Corollaries of Theorem 66 and Theorem 67: $\dfrac{-a}{b} = \dfrac{a}{-b}, \ b \neq 0. \ \dfrac{a}{-b} = \dfrac{-a}{b}, \ b \neq 0.$

Theorem 68

$$\frac{a}{b} + \frac{c}{d} = \frac{ad + bc}{bd} \quad (b \neq 0,\, d \neq 0).$$

Proof: Optional exercise.

Theorem 69

$$\frac{a}{b} - \frac{c}{d} = \frac{ad - bc}{bd} \quad (b \neq 0,\, d \neq 0).$$

Proof: Optional exercise.

Theorem 70

$$\frac{a}{b} \cdot \frac{c}{d} = \frac{ac}{bd} \quad (b \neq 0,\, d \neq 0).$$

Proof: Optional exercise.

Corollary of Theorem 70: $\dfrac{1}{b} \cdot a = \dfrac{a}{b} \quad (b \neq 0).$

Theorem 71

$$\frac{a}{b} \div \frac{c}{d} = \frac{ad}{bc} \quad (b \neq 0,\, c \neq 0,\, d \neq 0).$$

Proof: Optional exercise.

The reader certainly knows the very basics of powers and roots of real numbers.

The *number* $a^n = a \cdot a \cdot a \cdot \ldots \cdot a$ (a real $\neq 0$, $n = 0, 1, 2, 3, \ldots$) is called the **nth power of a**, or **a to the nth power**. a is called the **basis** of the power, and n is called the **exponent** of the power.

If n is 1, 3, 5, … (positive odd interger), then for each real a there is exactly *one* real b such that $b^n = a$. This b is called the *n*th root of a and is denoted by $\sqrt[n]{a}$ or $a^{\frac{1}{n}}$. If $n = 2, 4, 6, \ldots$

28

(even interger) the situation is different. In this case, if a is negative, there is *no* real b such that $b^n = a$ because $b^n \geq 0$ for all real b. However, if a is positive it can be shown that there is *one and only one* positive b such that $b^n = a$. This *number b* is called the *positive nth root of a* and is denoted by $b = \sqrt[n]{a}$ or $b = a^{\frac{1}{n}}$. Since n is even $(-b)^n = b^n$ and hence each $a > 0$ has two real nth roots, namely b and $-b$. The *number $-b$* is called the *negative nth root of a* and is denoted by $b = -\sqrt[n]{a}$ or $b = -a^{\frac{1}{n}}$. Note: The symbols $\sqrt[n]{a}$ and $a^{\frac{1}{n}}$ are reserved for the *positive n*th root of *a*.

The *extraction of roots* is to be regarded as an inversion of *exponentiation*. Within the domain of positive numbers, extraction of roots and exponentiation are operations inverse to each other. The expression $\sqrt[n]{a}$ is called the **radical**. The number a from which the roots are to be extracted is called the **radicand**, the value a corresponds to the **basis** of the power, and the term **exponent** is also used here for n.

If n = 2 (and $a > 0$), then $b = \sqrt[2]{a} = \sqrt{a}$ is called the **positive square root** of a and $b = -\sqrt{a}$ is called the **negative square root** of a. If n = 3, then $b = \sqrt[3]{a}$ is called the **cube root** of a.

Note: $(+1)^n = +1$, $(-1)^{2n} = +1$, $(-1)^{2n-1} = -1$, where $n = 0, 1, 2, 3, \ldots$.

Powers and roots with *real exponents* can be introduced through Axioms 0 – 16 above and limiting processes (see a calculus textbook if you are interested).

A *p*th (where p is a real number $\neq 0$) root of a *positive real* number a is a *real* number b satisfying $b^p = a$ and written $b = \sqrt[p]{a}$. (See Theorem 73.) For $p = 2, 4, 6, 8, \ldots$ positive real numbers have one positive and one negative *p*th root. For $p = 1, 3, 5, 7, \ldots$ positive real numbers have one positive *p*th root but no negative real root. Negative real numbers do not have a *real p*th root if $p = 2, 4, 6, 8, \ldots$. (See Chapter III.)

Note that, in the *complex-number system* (see Chapter III), in general, the symbol $\sqrt[p]{a}$ (where a is a *real* number, $p \neq 0$) is *not* unique (that is b is not a number, b is a *set* of numbers), and some or all of the roots of a given *real* number a may *not* be real numbers (they may be complex numbers).

If p is 1, 2, 3, 4, ... , then, *in the complex-number system*, a has 1, 2, 3, 4, ... roots, respectively, if each root is counted up to its *multiplicity*. Remember that the real numbers are a *subset* of the set of complex numbers. (See Chapter III.)

Note: In the theorems below the numbers a, b, c are positive real numbers because we are here generally operating in the *real number system*.

Theorem 72

$(a^p)^q = a^{pq}$ (a real number > 0, p is a real number, q is a real number).

Proof omitted. But remember that powers and roots with real exponents can be introduced through Axioms 0 – 16 above and limiting processes (if interested, see a calculus textbook).

Examples:

$$((3.1)^\pi)^{\frac{2}{3}} = (3.1)^{\frac{2\pi}{3}} = \sqrt[3]{(3.1)^{2\pi}} = 3.1138...$$

$$((-3.1)^\pi)^{\frac{2}{3}} = (-3.1)^{\frac{2\pi}{3}} = \sqrt[3]{(-3.1)^{2\pi}} = 3.1138...$$ Note: Here $a < 0$. You see, all the theorems here work (with correct modifications) in the complex-number system, not only in the real-number system. See Chapter III.

Theorem 73

$a^{\frac{p}{q}} = \sqrt[q]{a^p}$ (p is a real number, q is a real number $\neq 0$, a is a real number > 0). Note: Why a > 0? Here is why: Example: $^{(-3)}\!\sqrt{0} = {}^{(-3)}\!\sqrt{0^1} = 0^{\frac{1}{-3}} = 0^{\frac{-1}{3}} = \sqrt[3]{0^{-1}} = \sqrt[3]{\frac{1}{0^1}} = \sqrt[3]{\frac{1}{0}}$ is not a number.

See theorems below.

Proof:

$$\left(a^{\frac{p}{q}}\right)^q = a^{\frac{p}{q}q} = a^p \rightarrow a^{\frac{p}{q}} = \sqrt[q]{a^p}.$$

Examples (p real $\neq 0$, q real $\neq 0$, a real > 0):

$$a^{\frac{1}{1}} = \sqrt[1]{a^1} = \sqrt[1]{a} = a$$

$$a^{\frac{1}{2}} = \sqrt[2]{a^1} = \sqrt{a}$$

$$a^{\frac{1}{p}} = \sqrt[p]{a}$$

$$a^{\frac{q}{p}} = a^{\frac{1}{\frac{p}{q}}} = \sqrt[\frac{p}{q}]{a}$$

$$(-\pi)^{\frac{8.0}{-0.66}} = \sqrt[-0.66]{(-\pi)^{8.0}} = 0.00000094176...$$

$$a^{\pi} \approx a^{\frac{314}{100}} = \sqrt[100]{a^{314}} .$$

Note:

$\sqrt[p]{0}$, where p $\leq$ 0, is not a number. (See explanation above.)

$\sqrt[0]{a}$ is not a number. (Formally $\sqrt[0]{a} = a^{\frac{1}{0}}$. See theorem above and below.)

0^0 is not a number. (Formally $0^0 = 0^{0 \cdot (-1)} = 0^{(-1) \cdot 0} = \left(0^{(-1)}\right)^0 = \left(\dfrac{1}{0^1}\right)^0$. See theorem below.)

0^{-1} is not a number. (Formally $0^{-1} = \dfrac{1}{0^1}$. See theorem below.)

$0^1 = 0$ (See theorem below.)
$0^p = 0$ (See theorem below.)
$1^a = 1$ (a is a real number.)
$a^0 = 1$ ($a \neq 0$) (See theorem below.)
$a^1 = a$ (See theorem below.)
$\sqrt[1]{a} = a$ (See theorem below.)
$\sqrt[p]{1} = 1$ (See theorem below.)

Theorem 74
$a^p \cdot a^q = a^{p+q}$ (a is a real number > 0, p is a real number, q is a real number).

Proof omitted. But remember that powers and roots with real exponents can be introduced through Axioms 0 – 16 above and limiting processes (if interested, see a calculus textbook).

31

Examples:

$$a^p \cdot a^{-p} = a^{p+(-p)} = a^0 = 1 \quad \text{(See theorem below.)}$$

$$a^{5.3} \cdot a^{-9.8} = a^{5.3-9.8} = a^{-4.5} = a^{\frac{-45}{10}} = \sqrt[10]{a^{-45}} = \sqrt[10]{\frac{1}{a^{45}}} \quad \text{(see theorem below).}$$

Theorem 75

$a^p \div a^q = a^{p-q}$ (a is a real number > 0, p is a real number, q is a real number).

Proof:

$$a^q \cdot a^{p-q} = a^{q+(p-q)} = a^p.$$

Examples:

$$\pi^\pi \div \pi^\pi = \pi^{\pi-\pi} = \pi^0 = 1. \quad \text{(See theorem below.)}$$

$$\frac{a^p}{a^q} = a^{p-q}.$$

Theorem 76

$(ab)^p = a^p \cdot b^p$ (a is a real number > 0, b is a real number > 0, p is a real number, q is a real number).

Proof omitted. But remember that powers and roots with real exponents can be introduced through Axioms 0 – 16 above and limiting processes (if interested, see a calculus textbook).

Example:

$$(2\pi)^3 = 2^3 \pi^3 = 8\pi^3.$$

Theorem 77

$\left(a \div b\right)^{p} = a^{p} \div b^{p}$ (a is a real number > 0, b is a real number > 0, p is a real number, q is a real number).

Proof:

$$b^{p} \cdot \left(a \div b\right)^{p} = \left[b\left(a \div b\right)\right]^{p} = a^{p}.$$

Examples:

$$\left(0.1 \div 3.6\right)^{-2.9} = 0.1^{-2.9} \div 3.6^{-2.9} \approx 32604$$

$$\left(\frac{a}{b}\right)^{p} = \frac{a^{p}}{b^{p}}.$$

Theorem 78

$\sqrt[p]{ab} = \sqrt[p]{a}\sqrt[p]{b}$ (a and b are real numbers > 0, p is a real number $\neq 0$).

Proof:

$$\sqrt[p]{a}\sqrt[p]{b} = a^{\frac{1}{p}}b^{\frac{1}{p}} = \left(ab\right)^{\frac{1}{p}} = \sqrt[p]{ab}.$$

Theorem 79

$a^{1} = a$ (a is a real number).

Proof: Optional exercise (use $a > 0$).

Theorem 80

$\sqrt[1]{a} = a$ (a is a real number).

Proof: Optional exercise (use $a > 0$).

Theorem 81

$a^0 = 1$ (a is a real number $\neq 0$).

Proof: Optional exercise (use $a > 0$).

Theorem 82

$0^p = 0$ (p is a real number > 0).

Proof: Optional exercise.

Theorem 83

$$a^{-p} = \frac{1}{a^p}$$ (a is a real number > 0, p is a real number).

Proof: Optional exercise.

Theorem 84

$\sqrt[p]{1} = 1$ (p is a real number $\neq 0$).

Proof: Optional exercise.

Theorem 85 (Generalization of Theorem 5)

$(abc...)^p = a^p b^p c^p ...$ (a, b, c are real numbers > 0, p is a real number).

Proof: Optional exercise.

Theorem 86 (Generalization of Theorem 7)

$\sqrt[p]{abc...} = \sqrt[p]{a}\sqrt[p]{b}\sqrt[p]{c}...$ (a, b, c are real numbers > 0, p is a real number $\neq 0$).

Proof: Optional exercise.

Theorem 87

$$a^p = \sqrt[q]{\left(a^p\right)^q} \quad (a \text{ is a real number} > 0, p \text{ is a real number}, q \text{ is a real number} \ne 0).$$

Proof: Optional exercise.

Theorem 88

$$\left(\sqrt[p]{a}\right)^q = \sqrt[p]{a^q} \quad (a \text{ is a real number} > 0, p \text{ is a real number} \ne 0, q \text{ is a real number}).$$

Proof: Optional exercise.

Theorem 89

$$\sqrt[p]{\sqrt[q]{a}} = \sqrt[pq]{a} \quad (a \text{ is a real number} > 0, p \text{ is a real number} \ne 0, q \text{ is a real number} \ne 0).$$

Proof: Optional exercise.

Theorem 90

$$\sqrt[p]{a} = \sqrt[qp]{a^q} \quad (a \text{ is a real number} > 0, p \text{ is a real number} \ne 0, q \text{ is a real number} \ne 0).$$

Proof: Optional exercise.

Theorem 91

$$\sqrt[p]{a} = \sqrt[p+q]{\sqrt[q]{a}} \quad (a \text{ is a real number} > 0, p \text{ is a real number} \ne 0, q \text{ is a real number} \ne 0).$$

Proof: Optional exercise.

Theorem 92

If $a \ne 0$, b and c are real numbers, and $ax^2 + bx + c = 0$, then $x = \dfrac{-b \pm \sqrt{b^2 - 4ac}}{2a}$.

Let $x_1 = \dfrac{-b + \sqrt{b^2 - 4ac}}{2a}$ and $x_2 = \dfrac{-b - \sqrt{b^2 - 4ac}}{2a}$. Then $x_1 + x_2 = -\dfrac{b}{a}$ and $x_1 \cdot x_2 = \dfrac{c}{a}$.

35

Note: If $b^2 - 4ac \geq 0$ then x is real, and if $b^2 - 4ac < 0$ then x is complex. (See Chapter III.)

Proof: Optional exercise.

Optional exercises:

Use Theorem 92 to show that if:

$x^2 = 4$, then $x = \pm 2$,

$x^2 = -x$, then $x_1 = 0$ and $x_2 = -1$,

$x^2 + 4 = 0$, then, $x = \pm\sqrt{-4}\ (=\pm 2i)$,

$x^2 + 2x + 1 = 0$, then $x_1 = -1$ and $x_2 = -1$.

Solving irrational equations:

"I'm not familiar with this term [irrational equations]."
Doctor Rick

Any equation where the *varable* is *inside a radical* is called an **irrational equation** or **radical** equation. When solving an irrational equation, the key step will be removing the radical by raising expressions to powers. Raising expressions to powers can lead to **extraneous solutions** (**roots**) (additional solutions). We *must* often check which values are **true solutions** (see below).

Given $a = b$, where a is a real number and n is a real number, then $a = b \rightarrow a^n = b^n$ (the Principle of Power theorem).

If n is an *even integer*, then $(-a)^n = b^n \rightarrow a^n = b^n$. In this case $a^n = b^n$ represents two different equations, namely $a = b$ and $-a = b$. Here we *must* check which values are *true solutions*.

Examples:

$$\sqrt{6x+10} = -2x \rightarrow \left(\sqrt{6x+10}\right)^2 = (-2x)^2 \rightarrow 6x+10 = 4x^2 \rightarrow 4x^2 - 6x - 10 = 0$$

$$\rightarrow 2\left(2x^2 - 3x - 5\right) = 0 \rightarrow 2(x+1)(2x-5) = 0 \rightarrow x = -1 \text{ (true) or } x = \frac{5}{2} \text{ (not true)}.$$

Note: Here by *convention* (*definition*) $\sqrt{6x+10} = \sqrt{25} = 5$ is a *number* (and therefore is *unique*), namely the **principal value** of $\sqrt{25}$ or the *positive square root* of 25. The other square root of 25, namely -5, is the *negative square root* of 25. If we, in the equation above, used the *set of numbers* $\sqrt{6x+10} = \sqrt{25} = \{5, -5\}$ (see Chapter III) instead of the *number*

$\sqrt{6x+10} = \sqrt{25} = 5$ then $x = \dfrac{5}{2}$ too would be a *true solution*.

$$\sqrt{x} = x-2 \rightarrow \left(\sqrt{x}\right)^2 = (x-2)^2 \rightarrow x = x^2 - 4x + 4 \rightarrow x_1 = 4, \ x_2 = 1.$$
True solution: $x = 4$.

$$\sqrt[3]{2x-5} = 3 \rightarrow (2x-5)^{\frac{1}{3}} = 3 \rightarrow \left[(2x-5)^{\frac{1}{3}}\right]^3 = 3^3 \rightarrow 2x-5 = 27 \rightarrow x = 16 \text{ (is the only}$$

solution (root)).

$$\sqrt{x}+1 = \sqrt{x+9} \rightarrow x + 2\sqrt{x} + 1 = x + 9 \rightarrow 2\sqrt{x} + 1 = 9 \rightarrow \sqrt{x} = 4 \rightarrow x = 16 \text{ (because } 4 \cdot 4 =$$
x) (is the only solution).

Remember that the **absolute value** of a real number a is denoted by $|a|$, and $|a| = a$ if $a \geq 0$ and $|a| = -a$ if $a < 0$. Therefore $\sqrt{a^2} = |a|$, where a is a real number. So we have: $(x+5)^2 = 18 \ \rightarrow \ |x+5| = 3\sqrt{2} \ \rightarrow \ x = 3\sqrt{2} - 5 \text{ (true) or } x = -3\sqrt{5} - 5 \text{ (true)}.$

A basic **radical equation** is an equation of the form $x^{\frac{m}{n}} = a$, for m, n integers $\neq 0$ and a a real number > 0, which has solution $x = \sqrt[m]{a^n} = \left(\sqrt[m]{a}\right)^n$ if m is *odd*, and solution

$x = \pm\sqrt[m]{a^n} = \pm\left(\sqrt[m]{a}\right)^n$ if m is *even*.

Example: $(x+5)^{\frac{2}{3}} = 4 \rightarrow x+5 = \pm\sqrt[3]{4^2} \rightarrow x_1 = 3, x_2 = -13$.

Solving linear equations:

A ***linear equation*** is an equation in which each term is either a constant or the product of a constant and (the first power of) a single variable. Linear equations will always have one solution, no solutions, or an infinite number of solutions. A linear ***determinate equation*** has one solution. A ***contradictory equation*** has no solution. An ***indeterminate equation*** possesses an infinite number of solutions.

The eight *equation* $\leftrightarrow$ *solution* below are, of course, eight *theorems* of real numbers.

$$x + B = C \leftrightarrow x = C - B$$

$$x - B = C \leftrightarrow x = C + B$$

$$Ax = C \leftrightarrow x = \frac{C}{A} \quad (A \neq 0)$$

$$\frac{x}{A} = C \leftrightarrow x = C \cdot A \quad (A \neq 0)$$

$$Ax + B = C \leftrightarrow x = \frac{C - B}{A} \quad (A \neq 0)$$

$$Ax - B = C \leftrightarrow x = \frac{C + B}{A} \quad (A \neq 0)$$

$$\frac{x}{A} + B = C \leftrightarrow x = (C - B)A \quad (A \neq 0)$$

$$\frac{x}{A} - B = C \leftrightarrow x = (C + B)A \quad (A \neq 0).$$

Optional exercise: Check the solutions of the linear equations above.

Solving a system of linear equations in the variables (unknowns) *x* and *y*:

Given

$$a_1 x + b_1 y + c_1 = 0$$
$$a_2 x + b_2 y + c_2 = 0,$$

where $a_1, b_1, c_1, a_2, b_2, c_2$ are real constants, then

$$x = -\frac{c_1 b_2 - c_2 b_1}{a_1 b_2 - a_2 b_1}$$

$$y = -\frac{a_1 c_2 - a_2 c_1}{a_1 b_2 - a_2 b_1}$$

$$(a_1 b_2 - a_2 b_1 \neq 0).$$

Optional exercise: Check the theorem above, i.e. check the solutions of the system of linear equations above.

The ordinary real space of three dimensions: Coordinate system, point, plane, straight line

In the ordinary real *space* of three dimensions a *point* is equivalent to an ordered set of three real numbers (x, y, z) called the *coordinates* of the point. The totality of all such points with coordinates (x, y, z) constitute the ordinary real space of three dimensions. (If x, y, z are any complex numbers, we are dealing with the ordinary complex space of three dimensions.) In this space we introduce a *rectangular coordinate system* with mutually perpendicular *coordinate axes* (*x*-axis, *y*-axis, *z*-axis) passing through the *origin* with coordinates $(0, 0, 0)$. In mathematics we are mostly using *right-handed* rectangular coordinate systems. A right-handed rectangular coordinate system in 3-space has the property that an *ordinary screw* pointed in the positive direction of the *z*-axis would be *advanced* if the positive *x*-axis is rotated $90°$ toward the positive *y*-axis. (The system is *left-handed* if the screw would be *retracted*.)

A *plane* (flat surface) in this space is the set of all points (x, y, z) which satisfy a single linear equation

$$ax + by + cz + d = 0,$$

where a, b, c, d are real constants and a, b, c are not all zero.

39

Optional proof: Hint: Use vector calculus to prove this theorem.

A *straight line* (in the ordinary real space) is the set of all points (x, y, z) which satisfy two linear equations

$$a_1 x + b_1 y + c_1 z + d_1 = 0$$
$$a_2 x + b_2 y + c_2 z + d_2 = 0,$$

where $a_1, b_1, c_1, a_2, b_2, c_2$ are real constants, a_1, b_1, c_1 are not all zero, and a_2, b_2, c_2 are not all zero, provided the relations $\dfrac{a_1}{a_2} = \dfrac{b_1}{b_2} = \dfrac{c_1}{c_2}$ do not hold.

Optional proof: Hint: Use vector calculus to prove this theorem.

Optional exercises:

What is the equation of the yz-plane?
Answer: $x = 0$.

What is the equation of the xz-plane?
Answer: $y = 0$.

What is the equation of the xy-plane?
Answer: $z = 0$.

What are the equations of the x-axis?
Answer: $y = 0$, $z = 0$.

What are the equations of the y-axis?
Answer: $x = 0$, $z = 0$.

What are the equations of the z-axis?
Answer: $x = 0$, $y = 0$.

What are the equations of the origin?
Answer: $x = 0$, $y = 0$, $z = 0$.

What are the equations of the point (a, b, c)?
Answer: $x = a, y = b, z = c$.

A *straight line* in the xy-plane has equation $Ax + By + C = 0$ (Why?), where the real constants A and B are not both equal to zero.

What is the equation of a *straight line* in the xz-plane?
Answer: $Ax + Bz + C = 0$.

What is the equation of a *straight line* in the yz-plane?
Answer: $Ay + Bz + C = 0$.

Flat and curved spaces

It has been proved that the angles of an ordinary triangle in an ordinary three dimensional space (an Euclidean 3-space) sum to 180 degrees. Therefore Euclidean 3-space is said to be *flat*. The surfaces of an ordinary cylinder and an ordinary cone are Euclidean 2-spaces, and are flat in the sense of our "definition". But the angles of a triangle on the surface of a sphere (a two dimensional space) do not sum to 180 degrees, and therefore the surface of a sphere is not flat. A space which is not flat is called *curved*.

F. More optional exercises (with answers):

1. Prove that $a + b - b = a$.

2. Prove that $a + \dfrac{b}{c} = \dfrac{ac + b}{c}$ $(c \neq 0)$.

3. Prove that $n(a + b + c) = na + nb + nc$.

4. Prove that $x^2 + (a + b)x + ab = (x + a)(x + b)$.

5. Prove that $a \div b = a \cdot \dfrac{1}{b}$ $(b \neq 0)$.

6. Prove that $\dfrac{a}{b} \div t = \dfrac{a}{bt}$ $(b \neq 0,\ t = 0)$.

7. Prove that $\dfrac{a}{b} \div t = \dfrac{a \div t}{b}$ $(b \neq 0,\ t \neq 0)$.

41

8. Prove that $\dfrac{a}{b} = \dfrac{\frac{a}{c}}{\frac{b}{c}}$ $(b \neq 0,\ c \neq 0)$.

9. Prove that $\dfrac{a}{b} = c \leftrightarrow \dfrac{b}{a} = \dfrac{1}{c}$ $(b \neq 0,\ a \neq 0,\ c \neq 0)$.

10. Prove that $\dfrac{\frac{1}{b}}{a} = \dfrac{a}{b}$ $(a \neq 0,\ b \neq 0)$.

11. Prove that $\dfrac{\frac{1}{a}}{\frac{1}{b}} = \dfrac{b}{a}$ $(a \neq 0,\ b \neq 0)$.

12. Prove that $\dfrac{a}{b} = \dfrac{c}{d} \leftrightarrow \dfrac{b}{a} = \dfrac{d}{c}$ $(a,b,c,d \neq 0)$.

13. Prove that $\dfrac{a}{b} = \dfrac{c}{d} \leftrightarrow \dfrac{a}{c} = \dfrac{b}{d}$ $(b,c,d \neq 0)$.

14. Prove that $\dfrac{a}{b} = \dfrac{c}{d} \leftrightarrow \dfrac{a \pm b}{b} = \dfrac{c \pm d}{d}$ $(b \neq 0,\ d \neq 0)$.

15. Prove that $\dfrac{a}{b} = \dfrac{c}{d} = \dfrac{e}{f} \leftrightarrow \dfrac{a+c+e}{b+d+f} = \dfrac{a}{b}$ $(b,d,f \neq 0)$.

16. Prove that $\dfrac{a}{b} = \dfrac{c}{d}$ and $a = c \leftrightarrow b = d$ $(b \neq 0,\ d = 0)$.

17. Show that $\dfrac{a}{\sqrt{a}} = \sqrt{a}$ $(a > 0)$.

18. Show that $\dfrac{\sqrt{15}}{\sqrt{3}} = \sqrt{5}$.

19. Show that $\dfrac{\sqrt{2}}{\sqrt{3}} = \dfrac{\sqrt{6}}{3}$.

20. Show that $\sqrt{x^2} = |x|$, where x is a real number.

21. Show that $\left(\sqrt{x}\right)^2 = x$, where x is a real number ≥ 0.

22. Show that $x^2 \geq 0$, where x is a real number.

23. Show that $\dfrac{x}{\sqrt{x}} = \sqrt{x}$, where $x > 0$.

24. Show that if $x^2 = a^2$ $(a \neq 0)$, then $x = \pm a$.

25. Show that $\dfrac{\left(\sqrt[4]{x}\right)^2 \cdot \sqrt[3]{y^{-4}}}{\left(\sqrt[8]{x}\right)^3 \cdot \sqrt{y^{-1}}} = \dfrac{x^{\frac{1}{8}}}{y^{\frac{5}{6}}}$.

26. Prove that $\sqrt[-n]{a} = \dfrac{1}{\sqrt[n]{a}}$, where $a > 0$, and $n = 1, 2, 3, \ldots$.

27. Prove that $\sqrt[-n]{a} = \dfrac{1}{\sqrt[n]{a}}$, where $a < 0$, and $n = 1, 3, 5, \ldots$.

28. Using the formula $\dfrac{a^n - b^n}{a - b} = a^{n-1} + a^{n-2}b + a^{n-3}b^2 + \ldots + ab^{n-2} + b^{n-1}$, where $n = 1, 2,$

3, $\ldots$, show that $\dfrac{a^3 - b^3}{a - b} = a^2 + 2ab + b^2$.

29. Solve the following system of equations:
$$x + 2y = 3,$$
$$4x + 5y = 6.$$
Answer: $x = -1, y = 2$.

III THE COMPLEX-NUMBER SYSTEM

A. Discovering a new axiom and new numbers

The axioms are children of *experience*. Actually, we do not really *understand* the axioms. The axioms are *discovered*. That's it. That's all. The theorems, in turn, are *proved* by using the axioms and nothing else. (The rules of logic are axioms, too.) So, stop pondering about a proved mathematical result. If it is proved, it is proved. Example: We can not, and shall not, directly understand results like $\left(-\sqrt{1}\right)\left(-\sqrt{1}\right) = +1$, or $\left(\sqrt{-1}\right)\left(\sqrt{-1}\right) = -1$, or $e^{\pi i} = -1$, or ... . Proof! That's the thing about mathematics.

If $a > 0$ and $n = 2, 4, 6, 8, \ldots$ then the number $\sqrt[n]{-a}$ is not a real number. The real-number system is not mathematically complete. So, in the real-number system not every **polynomial equation** $c_n x^n + c_{n-1} x^{n-1} + c_{n-2} x^{n-2} + \ldots + c_2 x^2 + c_1 x + c_0 = 0$ has a **real root (real solution)**.

Calculations, *following the axioms and theorems of real numbers*: If $x^2 + 1 = 0$ then $x = \pm\sqrt{-1} = \pm$ **i**, where $\mathbf{i} = \sqrt{-1}$ and $\mathbf{i}^2 = -1$. **i** is called the *imaginary unit*. If $x^2 + px + q = 0$, and p, q are real numbers, then we know that $x = -\dfrac{p}{2} \pm \sqrt{\left(\dfrac{p}{2}\right)^2 - q}$, and that if $\left(\dfrac{p}{2}\right)^2 - q < 0$, then

there is no real solution. However, $x = -\dfrac{p}{2} \pm \sqrt{\left(\dfrac{p}{2}\right)^2 - q} = -\dfrac{p}{2} \pm \sqrt{\left[q - \left(\dfrac{p}{2}\right)^2\right](-1)}$

$= -\dfrac{p}{2} \pm \sqrt{\left[q - \left(\dfrac{p}{2}\right)^2\right]}\sqrt{-1} = -\dfrac{p}{2} \pm \left(\sqrt{q - \left(\dfrac{p}{2}\right)^2}\right)i = a + bi$, where $a = -\dfrac{p}{2}$ and

$b = \pm\sqrt{q - \left(\dfrac{p}{2}\right)^2}$ are real numbers and **i** is the imaginary unit, mentioned above. Result: We have *calculated* and *discovered* a new kind of numbers, namely $z = a + b\mathbf{i}$, which are real if $b = 0$. The numbers $z = a + b\mathbf{i}$, called **complex numbers**, include the real numbers and follow the axioms (and consequently also the theorems) of real numbers (as shown below). $b\mathbf{i}$ is called a **pure imaginary number.**

By convention (definition), the *principal square root* (see later) of -1 is $\mathbf{i}$, or more generally, if a is any positive real number, then the *principal square root* of $-a$ is $\sqrt{-a} = \left(\sqrt{a}\right)i$. Proof:

$$\left[\left(\sqrt{a}\right)i\right]\left[\left(\sqrt{a}\right)i\right] = -a.$$

The *discovered* and *calculated* $\mathbf{i}$ is the **imaginary unit** just like 1 is the *real unit*. (Electrical engineers use $\mathbf{j}$ because $\mathbf{i}$ is used to mean *current* in their equations.) Note that here we are following the rules (axioms and theorems) of real numbers. Complex numbers (see later) do not bend the rules of real numbers. On the contrary, complex numbers follow the rules of real numbers. Examples: If z and w are complex numbers we have $(-z)(-w) = +(zw)$, $(-i)(-i) = +(i \cdot i)$, $\sqrt{z}\sqrt{z} = z$, etc. But, as shown later, *complex numbers, including the imaginary unit $\mathbf{i}$, is neither positive nor negative. Therefore, $\mathbf{i} = \sqrt{-1}$ and $\mathbf{i}^2 = -1$, do not bend the rules. On the contrary, $\mathbf{i} = \sqrt{-1}$ and $\mathbf{i}^2 = -1$ follow from the rules.* By the way, note, for example: $\sqrt{5}\sqrt{5} = 5$, $\sqrt{1}\sqrt{1} = 1$, $\sqrt{0}\sqrt{0} = 0$, $\sqrt{-1}\sqrt{-1} = -1$, $\sqrt{-5}\sqrt{-5} = -5$, etc. Beautiful!

Let $\mathbf{i} = \sqrt{-1}$ and $z = a + b\mathbf{i}$, where a and b are real numbers, be **Axiom 17**. (Note that this new axiom is a **partly dependent** axiom.) And let the theory of limiting processes (mentioned earlier) be **Axiom 00**. Then we now have totally 19 axioms: Axiom 00, Axiom 0, Axiom 1, Axiom 2, ... , and Axiom 17.

The form $z = a + b\mathbf{i}$ is called the **standard form, Cartesian form, rectangular form,** or **algebraic form** of the complex number z. (The form $z = (a, b)$ (see below) is an ordered pair of complex numbers.)

It's discovered and proved (see below) that the 19 axioms above are creating an **extension** of the real-number system. The new and mathematically complete number system is the **complex-number system. The Fundamental Theorem of Algebra (FTA)** states that every polynomial equation of degree n with complex coefficients has n roots in the complex-number system if each solution (root) is counted up to its multiplicity. Example: If $z^4 - 4z + 3 = 0$, then $z_1 = 1$, $z_2 = 1$, $z_3 = -1 + \left(\sqrt{2}\right)i$, $z_4 = -1 - \left(\sqrt{2}\right)i$.

The imaginary unit $\mathbf{i}$ is, of course, used to pull together the real-number system and the complex number system.

A complex number $z = a + b\mathbf{i}$ may be represented as a **point** with coordinates (a, b) in the **complex plane** where the x-axis is the **real line (real axis)** and the y-axis is the **imaginary line (imaginary axis)**. $z = a + b\mathbf{i}$ is obviously also the **position vector** of the point (or vector) (a, b). So a complex number $z = a + b\mathbf{i}$ may also be represented as a *position vector*. (The *complex plane* is also known as *Argand plane*, *Gauss plane*, or *z-plane*. It was first described by C. Wessel in a paper of 1797 (printed 1799). A plot of complex numbers in the plane is sometimes called an *Argand diagram*.) So, obviously, $z = (a, b) = a + b\mathbf{i}$. The form $z = (a, b)$ is an **ordered pair of real numbers.**

Remember: *Drawing useful figures is left as optional exercises.*

"There is nothing about the complex number plane to clue us in that $i^2 = \bullet 1$ or $\dfrac{1}{i} = -i$. Furthermore, if the complex number plane exhausted the meaning of imaginaries, then imaginaries would be completely redundant: all we would need is an extra spatial axis, which can be handled just fine without imaginaries. Unique properties of the complex number plane, like the $90°$ rotation of a number around the origin when multiplied by i, depend on the independent property that $i^2 = \bullet 1$." Isaac Asimov.

In the expression $z = a + b\mathbf{i}$ a is called the **real part** and b is called the **imaginary part**. Note: We sometimes write $z = \operatorname{Re} z + i \operatorname{Im} z$ where $\operatorname{Re} z = a$ and $\operatorname{Im} z = b$ (re = real, im = imaginary).

The lack of meaning of $\sqrt[n]{-a}$ or $(-a)^{\frac{1}{n}}$ $(a > 0, n = 2, 4, 6, 8, \ldots)$ in the *real-number system* was the basic problem with the system.

Axioms 00-17 imply the operations (multiplication, addition, division, subtraction) of complex numbers (as shown below), and the operations of complex numbers satisfy Axioms 00-17 (as shown below). Roots and powers are basically multiplication. So powers and roots of complex numbers do exist. Actually the symbol $\sqrt[n]{-a}$ $(a > 0, n = 2, 4, 6, 8, \ldots)$ has two meanings: representing the *set* of n complex numbers (see below) **or** representing the *complex number* $z = \sqrt[n]{-a}$, that is the *principal nth root* in the *set* $\sqrt[n]{-a}$ $(k = 0$ in the formula just below).

Optional exercise:

Check the correctness of the following formula (explanation and proof come later) for $n = 2$, $n = 3$, and $n = 4$:

$$\left\{ \left[\sqrt[n]{a} \cos\left(\frac{\pi}{n} + k\frac{2\pi}{n} \right) \right] + \left[\sqrt[n]{a} \sin\left(\frac{\pi}{n} + k\frac{2\pi}{n} \right) \right] i \right\}^n = -a$$

where $-\pi < \theta < \pi$ (see section III E) , a is a positive real number, $n = 2, 3, 4, 5, \ldots$, and $k = 0, 1, 2, 3, \ldots , (n-1)$.

(To find the nth roots of a complex number $w \neq 0$ we have to solve the equation $z^n = w$. This equation has n distinct solutions (roots), see later.)

Note that the formula shows: ***An imaginary number (except 0) is a number whose square is a negative number, or, equivalently, the square root of a negative number is an imaginary number***.

As stated above, Axioms 00-17 hold for complex numbers as well as real numbers (proofs come below). Therefore, ***the theorems of complex numbers are basically the same as the corresponding theorems of real numbers***. The complex numbers contain the real numbers. Therefore, ***the theorems which hold for complex numbers, are generally also valid for real numbers***.

Obviously, $a + b\mathbf{i} = c + d\mathbf{i}$ if and only if $a = c$ and $b = d$.

Optional exercise:

Prove that if $a + b\mathbf{i} = c + d\mathbf{i}$, then $a = c$ and $b = d$.

$z = a + 0\mathbf{i} = (a, 0) = a$ is the (pure) real number a. $z = 0 + 0\mathbf{i} = (0, 0) = 0$. $z = 1 + 0\mathbf{i} = (1, 0) = 1$. $z = 0 + 1\mathbf{i} = (0, 1) = \mathbf{i} = \sqrt{-1}$ is the imaginary unit. $z = 0 + b\mathbf{i} = (0, b) = b\mathbf{i}$ is a ***pure imaginary number***. Note that $(0, 0) = 0$ is also a pure imaginary number. $z = a + b\mathbf{i}$, where $b \neq 0$, is a ***non-real complex number***. $-z = -(a, b) = (-a, -b)$, see Theorem 93 below. By the way: $z = +z = +(a, b) = (+a, +b)$, see Theorem 94 below. Note: ***Such an ordering as > or < is generally impossible for complex numbers. Complex numbers are neither negative nor positive***.

If a and b are real and positive, then $\pm\sqrt{(-a)(-b)} = \sqrt{-a}\sqrt{-b}$. Proof:

$$\left(i^2\sqrt{ab} \right)^2 = i^2 \cdot a \cdot i^2 \cdot b \;\rightarrow\; i^2\sqrt{ab} = \pm\sqrt{i^2 \cdot a \cdot i^2 \cdot b} = \pm\sqrt{(-a)(-b)},$$

$$\sqrt{-a}\sqrt{-b} = \left[\left(\sqrt{a} \right)i \right]\left[\left(\sqrt{b} \right)i \right] = i^2\sqrt{ab} .$$ However, $\pm\sqrt{(-a)(-b)}$ is not a number, but a set

of numbers. Using numbers the theorem is simply $\sqrt{(-a)(-b)} = \sqrt{-a}\sqrt{-b}$, where $\sqrt{(-a)(-b)}$ is the *principal value* (see later) of $\pm\sqrt{(-a)(-b)}$. Similarly, if a and b are real and positive, then $\sqrt{(-a)(b)} = \sqrt{-a}\sqrt{b}$. Proof: Optional exercise. Using numbers (not sets of numbers), in the complex-number system we simply have $\sqrt{ab} = \sqrt{a}\sqrt{b}$ where a and b are real numbers. Using numbers (not sets of numbers) we also have $\sqrt{\dfrac{a}{b}} = \dfrac{\sqrt{a}}{\sqrt{b}}$ where a is real

and b is real $\neq 0$. Proof: Optional exercise. Examples: $\sqrt{\dfrac{-1}{1}} = \dfrac{\sqrt{-1}}{\sqrt{1}} = \dfrac{i}{1} = i$. Proof:

$i \cdot i = \dfrac{i}{1}\cdot\dfrac{i}{1} = \dfrac{-1}{1} = -1$. $\sqrt{\dfrac{1}{-1}} = \dfrac{\sqrt{1}}{\sqrt{-1}} = \dfrac{1}{i} = -i$. Proof: $(-i)(-i) = \dfrac{1}{i}\cdot\dfrac{1}{i} = \dfrac{1}{-1} = -1$.

More generally, using principal values only, $\sqrt{z_1 z_2} = \sqrt{z_1}\sqrt{z_2}$ where z_1 and z_2 are complex

numbers. And also (using numbers, that is using principal values only) $\sqrt{\dfrac{z_1}{z_2}} = \dfrac{\sqrt{z_1}}{\sqrt{z_2}}$ where

$z_2 \neq 0$. Proofs: Optional exercise.

Remember: ***The square of any imaginary number (except 0) is a negative number***. If $a > 0$ and $b \neq 0$, then:

$$\sqrt{-a}\sqrt{-a} = \sqrt{(-1)a}\sqrt{(-1)a} = \sqrt{-1}\sqrt{-1}\sqrt{a}\sqrt{a} = \left(\sqrt{a}\right)^2 i^2 = -a < 0$$

and $(bi)(bi) = b^2 i^2 = -b^2 < 0$.

Obviously, ***Axioms* 00, 1, 2, 3, 4, 5, 6, 7, *and* 17 *hold for complex numbers*. *But Axioms* 0, 8, 9, 10, 11, 12, 13, 14, 15, *and* 16 *must be proved to hold for complex numbers*.**

B. Axioms → operations

We do *not* have to *define* addition and multiplication of complex numbers. These operations follow from the axioms:

Addition of complex numbers follows from the axioms of complex (and real) numbers:

$$z + u = (a, b) + (c, d) = (a + b\mathbf{i}) + (c + d\mathbf{i}) = (a + c) + (b + d)\mathbf{i} = (a + c, b + d).$$

Subtraction of complex numbers follows from the axioms of complex (and real) numbers and addition of complex numbers:

$$z - u = (a, b) - (c, d) = (a + b\mathbf{i}) - (c + d\mathbf{i}) = (a - c) + (b - d)\mathbf{i} = (a - c, b - d).$$

Multiplication of complex numbers follows from the axioms of complex (and real) numbers:

$$zu = (a, b)(c, d) = (a + b\mathbf{i})(c + d\mathbf{i}) = (a + b\mathbf{i})c + (a + b\mathbf{i})d\mathbf{i} = (ac - bd) + (bc + ad)\mathbf{i} = (ac - bd, bc + ad).$$

Optional exercise: Prove that $k(a + b\mathbf{i}) = (ka) + (kb)\mathbf{i}$, where k is a real number.

Division of complex numbers follows from the axioms of complex (and real) numbers and multiplication of complex (and real) numbers:

$$\frac{(a,b)}{(c,d)} = \frac{a+bi}{c+di} = \frac{ac+bd}{c^2+d^2} + \frac{bd-ad}{c^2+d^2}i = \left(\frac{ac+bd}{c^2+d^2}, \frac{bd-ad}{c^2+d^2} \right).$$

Proof:

$$(c+di)\left(\frac{ac+bd}{c^2+d^2} + \frac{bd-ad}{c^2+d^2}i \right) = a+bi.$$

Note: There are free **complex number calculators** on the Internet.

C. Operations → axioms

Now, Axiom 0, 8, 9, 10, 11, 12, 13, 14, 15, and 16 must be proved to hold for complex numbers.

Obviously, Axiom 0 was implicitly proved in section B just above.

Axiom 8
$z + u = u + z.$

Proof:

$z + u = (a, b) + (c, d) = (a + c, b + d) = (c + a, d + b)$
$= (c, d) + (a, b) = u + z.$

Axiom 9
$z + (u + v) = (z + u) + v.$

Proof: Optional exercise.

Axiom 10
$zu = uz.$

Proof:

$zu = (a, b)(c, d) = (ac - bd, ad + bc) = (ca - db, da + cb)$
$= (ca - db, cb + da) = (c, d)(a, b) = uz.$

Axiom 11
$z(uv) = (zu)v.$

Proof:

$z(uv) = (a, b)[(c, d)(e, f)] = (a, b)(ce - df, cf + de)$
$= [a(ce - df) - b(cf + de), a(cf + de) + b(ce - df)]$
$= (ace - adf - bcf - bde, acf + ade + bce - bdf)$
$= (ace - bde - adf - bcf, acf - bdf + ade + bce)$
$= [(ac - bd)e - (ad + bc)f, (ac - bd)f + (ad + bc)e]$
$= (ac - bd, ad + bc)(e, f) = [(a, b)(c, d)](e, f) = (zu)v.$

Axiom 12
$z(u + v) = zu + zv.$

Proof:

$$z(u + v) = (a, b)[(c, d) + (e, f)] = (a, b)(c + e, d + f)$$
$$= [a(c + e) - b(d + f), a(d + f) + b(c + e)]$$
$$= (ac + ae - bd - bf, ad + af + bc + be)$$
$$= [(ac - bd) + (ae - bf), (ad + bc) + (af + be)]$$
$$= (ac - bd, ad + bc) + (ae - bf, af + be)$$
$$= (a, b) (c, d) + (a, b) (e, f) = zu + zv.$$

Axiom 13
$z + 0 = z$ and $1z = z.$

Proof: Optional exercise.

Axiom 14
$z + (-z) = 0.$

Proof:

$z + (-z) = z - z$ (Theorem 13, i.e. several other axioms)
$z - z = (a, b) - (a, b) = (a - a, b - b) = (0, 0) = 0$
$z + (-z) = 0$ (Axiom 5).

Axiom 15 (The law of existence of additive inverse)
(This is the same axiom as Axiom 14.)
For every complex number z there is a complex number u such that z + u = 0.

Axiom 16 (The law of existence of multiplicative inverse)
For every complex number z • 0 there is a complex number u such that zu = 1.

Proof:

$$\frac{1}{z} = \frac{(1,0)}{(a,b)} = \left(\frac{a}{a^2+b^2}, \frac{-b}{a^2+b^2}\right) \text{ and}$$

$$(a,b)\left(\frac{a}{a^2+b^2}, \frac{-b}{a^2+b^2}\right) = 1 \text{ or } z \cdot \frac{1}{z} = 1 \text{ or } z \cdot z^{-1} = 1$$

or $zu = 1$ where $u = z^{-1}$.

Note that here we used Theorem 79 and Theorem 83, among other theorems, i.e. we used several other axioms.

D. Some basic theorems

Theorem 93
•z = −(a, b) = (−a, −b).

Proof:

$-z = -(a, b) = -(a + b\mathbf{i})$
$-(a + b\mathbf{i}) = -a - b\mathbf{i}$ (Theorem 41)
$-a - b\mathbf{i} = -a + (-b\mathbf{i})$ (Theorem 13)
$-a + (-b\mathbf{i}) = -a + (-b)\mathbf{i}$ (Theorem 35)
$-a + (-b)\mathbf{i} = (-a, -b)$
$-z = -(a, b) = (-a, -b)$ (Axiom 5).

Theorem 94
+z = +(a, b) = (+a, +b).

Proof:

$+z = z$ (Axiom 6)
$z = (a, b)$
$(a, b) = (+a, +b)$ (Axiom 6)
$+z = (+a, +b)$ (Axiom5).

Theorem 95

$(-z)(-w) = zw.$

Proof:

$z = (a, b)$ and $w = (c, d)$ (given)
$(-z)(-w) = (-a, -b)(-c, -d)$ (Theorem 93)
$(-a, -b)(-c, -d) = (ac - bd, ad + bc)$ (Theorem 26)
$(ac - bd, ad + bc) = zw$ (Axiom 4)
$(-z)(-w) = zw$ (Axiom 5).

Theorem 96

$|z| = \sqrt{a^2 + b^2}$

Proof:

Remember that the **absolute value** of a real number a is denoted by $|a|$, and $|a| = a$ if $a \geq 0$, $|a| = -a$ if $a < 0$. The **absolute value** of a complex number $z = (a, b)$ is denoted by $|z|$, and $|z| = \sqrt{a^2 + b^2}$ Note: If $b = 0$ then $|z| = \sqrt{a^2} = |a|$. Geometrically, the absolute value of a complex number (and a real number) is the **distance** between the point (a, b) and $(0, 0)$ in the (complex) plane. Proof: Pythagoras' theorem. (If the number is real then $b = 0$.) The **distance** between any two points (real numbers) a and b on the real axis is $|a - b| = |b - a|$. The distance between any two points (complex numbers) z_1 and z_2 in the complex plane is $|z_1 - z_2| = |z_2 - z_1|$. Proof: Optional exercise.

Note: $|z|$ is also denoted by r and called the **modulus** of the complex number z.

Theorem 97

(The **conjugate** of the complex number $z = a + b\mathbf{i}$ is the complex number $\overline{z} = a - b\mathbf{i}$)

If $z = (a, b)$ then $z + \overline{z} = 2a$ or $a = \dfrac{z + \overline{z}}{2}$.

Proof: Optional exercise.

Theorem 98

(The *conjugate* of the complex number $z = a + bi$ is the complex number $\bar{z} = a - bi$)

If $z = (a, b)$ then $z - \bar{z} = 2bi$ or $b = \dfrac{z - \bar{z}}{2i}$.

Proof: Optional exercise.

Theorem 99

(The **conjugate** of the complex number $z = a + bi$ is the complex number $\bar{z} = a - bi$)

If $z = (a, b)$ then $z\bar{z} = |z|^2 = a^2 + b^2$.

Proof: Optional exercise.

Note that $\dfrac{z_1}{z_2} = \dfrac{z_1 \bar{z}_2}{z_2 \bar{z}_2} = \dfrac{z_1 \bar{z}_2}{|z_2|^2}$ where the denominator becomes the real number $|z_2|^2$.

Optional exercises:

Show that $\dfrac{1}{i} = -i$ and that $\dfrac{-1 + 3i}{2 - i} = -1 + i$ using the result just above.

Theorem 100

$$\overline{z + u} = \bar{z} + \bar{u}$$

$$\overline{z - u} = \bar{z} - \bar{u}$$

$$\overline{z \cdot u} = \bar{z} \cdot \bar{u}$$

$$\overline{\left(\dfrac{z}{u}\right)} = \dfrac{\bar{z}}{\bar{u}}.$$

Proof: Optional exercise.

Theorem 101

If a, b, and c are real numbers, $a \neq 0$, $b^2 - 4ac < 0$, and $az^2 + bz + c = 0$, then

$$z = -\frac{b}{2a} \pm \frac{\sqrt{4ac - b^2}}{2a} \mathbf{i}.$$

Proof: Optional exercise. Hint: If $a > 0$ then $\sqrt{-a} = \left(\sqrt{a}\right)\mathbf{i}$ and $b - c = -(c - b)$. Note that here $\sqrt{a}$ is single-valued because $\sqrt{a}$ is the imaginary part of a complex number.

More generally, the equation $uz^2 + vz + w = 0$, where u, v, w, and z are complex numbers, has the following solutions: $z = \dfrac{-v + \sqrt{v^2 - 4uw}}{2u}$, just as expected, since this theorem is an ordinary theorem that follows from our axioms. Note: Now the symbol $\sqrt{v^2 - 4uw}$ is 2-valued and denotes the two square roots of the complex number $v^2 - 4uw$.

Examples:

If $x^2 - 2x + 2 = 0$, then $z_1 = 1 + i$, and $z_2 = 1 - i$.

If $z^2 + 6z + 13 = 0$, then $z = -3 \pm 2i$.

If $z^2 - i = 0$, then $z = \pm \left[\dfrac{1}{\sqrt{2}}(1 + i)\right]$.

If $z^2 + (4 + 2i)z + (3 + 2i) = 0$, then $z_1 = -1$, and $z_2 = -3 - 2i$.

Note that for the first and second equation above we have $z_2 = \overline{z_1}$ and $z_1 = \overline{z_2}$. *The complex conjugate root theorem* states that if a polynomial in one variable with real coefficients has $z = a + bi$ as a root, then $\overline{z} = a - bi$ is also a root. One more example: $z^3 - 7z^2 + 41z - 87 = 0$ has roots (solutions): $z_1 = 2 + 5i$, $z_2 = 2 - 5i$, $z_3 = 3$. Then, as most readers probably know, we have

$$z^3 - 7z^2 + 41z - 87$$
$$= \left[z - (2 + 5i)\right] \cdot \left[z - (2 - 5i)\right] \cdot (z - 3)$$
$$= (z - 2 - 5i)(z - 2 + 5i)(z - 3)$$
$$= (z - 3)(z^2 - 4z + 29).$$

Note: In computing the product $(z-2-5i)(z-2+5i)$, the pure imaginary numbers cancel.

Generalization: The factorization of the left-hand side of the general polynomial equation is indirectly guaranteed by the Fundamental Theorem of Algebra:

$$z^n + a_1 z^{n-1} + \ldots + a_{n-1}z + a_n = (z - z_1)(z - z_2)\ldots(z - z_n).$$

Solving linear equations:

A ***linear equation*** is an equation in which each term is either a constant or the product of a constant and (the first power of) a single variable. Linear equations will always have one solution, no solutions, or an infinite number of solutions. A linear ***determinate equation*** has one solution. A ***contradictory equation*** has no solution. An ***indeterminate equation*** possesses an infinite number of solutions.

The eight *equation* $\leftrightarrow$ *solution* below are, of course, eight *theorems* of complex numbers (z, A, B, C are complex numbers).

$$z + B = C \leftrightarrow z = C - B$$

$$z - B = C \leftrightarrow z = C + B$$

$$Az = C \leftrightarrow z = \frac{C}{A} \quad (A \neq 0)$$

$$\frac{z}{A} = C \leftrightarrow z = C \cdot A \quad (A \neq 0)$$

$$Az + B = C \leftrightarrow z = \frac{C - B}{A} \quad (A \neq 0)$$

$$Az - B = C \leftrightarrow z = \frac{C + B}{A} \quad (A \neq 0)$$

$$\frac{z}{A} + B = C \leftrightarrow z = (C - B)A \quad (A \neq 0)$$

$$\frac{z}{A} - B = C \leftrightarrow z = (C + B)A \quad (A \neq 0).$$

Optional exercise: Check the solutions of the linear equations above.

Solving a system of linear equations in the variables (unknowns) z and w:

Given

$$a_1 z + b_1 w + c_1 = 0$$
$$a_2 z + b_2 w + c_2 = 0$$

where $a_1, b_1, c_1, a_2, b_2, c_2$ are complex constants, then

$$z = -\frac{c_1 b_2 - c_2 b_1}{a_1 b_2 - a_2 b_1}$$

$$w = -\frac{a_1 c_2 - a_2 c_1}{a_1 b_2 - a_2 b_1}$$

$(a_1 b_2 - a_2 b_1 \neq 0)$.

Optional exercise: Check the solutions of the system of linear equations above.

E Powers and roots

As explained in section A, the theorems of powers and roots of complex numbers are basically the same as the theorems of powers and roots of real numbers. Note that the theorems of powers and roots of complex numbers are working unlimited, that is we can operate with $\sqrt[n]{-a}$, where a is a real number > 0, and $n = 2, 4. 6, 8, \ldots$. By the way, remember: The Fundamental Theorem of Algebra (FTA) states that every polynomial equation of degree n with complex coefficients has n roots in the complex-number system if each solution (root) is counted up to its *multiplicity*. And remember: From the theorem $\sqrt[n]{zw} = \sqrt[n]{z}\sqrt[n]{w}$ (proved in APPENDIX 5), where z and w are complex numbers, and $n = 1$, 2, 3, $\ldots$ we have (as we already know) that the *complex number* $\sqrt{-a} = \sqrt{(-1)(a)}$ $= \sqrt{-1}\sqrt{a} = \left(\sqrt{a}\right)i$, where a is a real number > 0. But, again, note that the *set* of complex numbers $\sqrt{a}$ where a is a real number, is *multi-valued*, namely *2-valued*. But if we write the *complex number* $z = 3 + \left(\sqrt{9}\right)i$, say, then $\sqrt{9} = 3$ is a *real number* and so is *unique* (single-valued) by convention (definition); we are using the principal value here. So, note, the above $\sqrt{a}$

and $\left(\sqrt{a}\right)i$ are single-valued, i.e. are numbers. By the way: The theorems $(zw)^n = z^n w^n$ and $\sqrt[n]{zw} = \sqrt[n]{z}\sqrt[n]{w}$ are *directly* (and unnecessarily!) proved in APPENDIX 5.

Important: Remember that the terms (symbols) $\sqrt[n]{z} = z^{\frac{1}{n}}$ ($n = 1, 2, 3, \ldots$) have *two* different meanings: *numbers* or *sets of numbers*. As a *set of numbers*, this symbol is *n-valued*, but, of course, as a *number*, the symbol $\sqrt[n]{z}$ ($= z^{\frac{1}{n}}$) is *unique*. This *number* is the *principal value* (see later) of $\sqrt[n]{z}$ ($= z^{\frac{1}{n}}$) (*principal nth root of z*, see later). The symbol z^n ($n = 1, 2, 3, \ldots$) is single-valued (unique), but the general meaning (as a *set*) of the symbol z^w, where z and w are complex numbers, is *multi-valued*. However, as a *number* z^w is the *principal value* of z^w (see later).

Again, ***if we are operating in the real-number system then*** $\sqrt{x}$ ***is unique***. But, again, ***if we are operating in the complex-number system then the term (symbol)*** $\sqrt{z}$ ***as a set of complex numbers is not unique. However as a complex number*** $\sqrt{z}$ ***is unique***.

Caution: The complex theorem $\sqrt[n]{z}\sqrt[n]{w} = \sqrt[n]{zw}$ works, but note the following examples:

Numbers: $\quad\quad\quad\quad \sqrt{1}\sqrt{1} = 1.$
Sets of numbers: $\quad \sqrt{1}\sqrt{1} = \sqrt{1\cdot 1} = \sqrt{1} = \pm 1.$

Numbers: $\quad\quad\quad\quad \sqrt{1}\sqrt{-1} = 1i = i$, or $\sqrt{1}\sqrt{-1} = \sqrt{1\cdot(-1)} = \sqrt{-1} = i.$
Sets of numbers: $\quad \sqrt{1}\sqrt{-1} = \sqrt{1\cdot(-1)} = \sqrt{-1} = \pm i.$

Numbers: $\quad\quad\quad\quad \sqrt{-1}\sqrt{-1} = i\cdot i = -1$, or $\sqrt{-1}\sqrt{-1} = \sqrt{(-1)(-1)} = -1.$
Sets of numbers: $\quad \sqrt{-1}\sqrt{-1} = \sqrt{(-1)(-1)} = \sqrt{1} = \pm 1.$

Numbers: $\quad\quad\quad\quad \sqrt{4}\sqrt{4} = 4.$
Sets of numbers: $\quad \sqrt{4}\sqrt{4} = \sqrt{4\cdot 4} = \sqrt{16} = \pm 4$

Numbers:
$$\sqrt{4}\sqrt{-4} = 2\cdot 2i = 4i \text{, or}$$
$$\sqrt{4}\sqrt{-4} = \sqrt{4\cdot(-4)} = \sqrt{-16} = \sqrt{(-1)(4)(4)} = \sqrt{-1}\sqrt{(4)(4)} = i\cdot 4 = 4i$$

Sets of numbers:
$$\sqrt{4}\sqrt{-4} = \sqrt{4(-4)} = \sqrt{-16} = \pm 4i \,.$$

Numbers:
$$\sqrt{-4}\sqrt{-4} = 2i\cdot 2i = 4i^2 = -4 \text{, or } \sqrt{-4}\sqrt{-4} = \sqrt{(-4)(-4)} = -4 \,.$$

Sets of numbers:
$$\sqrt{-4}\sqrt{-4} = \sqrt{(-4)(-4)} = \sqrt{16} = \pm 4 \,.$$

Numbers:
$$\sqrt{4}\sqrt{-9} = 2\cdot 3i = 6i \text{, or}$$
$$\sqrt{4}\sqrt{-9} = \sqrt{4\cdot(-9)} = \sqrt{-36} = \sqrt{(-1)(6)(6)} = \sqrt{-1}\sqrt{6\cdot 6} = i\cdot 6 = 6i \,.$$

Sets of numbers:
$$\sqrt{4}\sqrt{-9} = \sqrt{4\cdot(-9)} = \sqrt{-36} = \pm 6i \,.$$

Numbers:
$$\sqrt{-i}\sqrt{-8i} = \left[\left(\frac{1}{\sqrt{2}} - \frac{1}{\sqrt{2}}i\right)\right]\left[(2-2i)\right] = \left(\sqrt{8}\right)i \,.$$

Sets of numbers:
$$\sqrt{-i}\sqrt{-8i} = \sqrt{\left(\frac{1}{\sqrt{2}} - \frac{1}{\sqrt{2}}i\right)\cdot(2-2i)} = \pm\left(\sqrt{8}\right)i \,.$$

Numbers:
$$\sqrt{i^4} = \sqrt{i^2 i^2} = \sqrt{(-1)(-1)} = -1 \text{ (note: } \sqrt{(-1)(-1)} \text{ is a number)}.$$

Sets of numbers:
$$\sqrt{i^4} = \sqrt{i^2 i^2} = \sqrt{(-1)(-1)} = \sqrt{1} = \pm 1 \,.$$

Numbers:
$$\sqrt{z}\sqrt{z} = z \text{ (and } \sqrt{z}\sqrt{z} = \sqrt{z\cdot z} = \sqrt{z^2} = z \text{)}.$$

Sets of numbers:
$$\sqrt{z}\sqrt{z} = \sqrt{z\cdot z} = \sqrt{z^2} = \pm z \,.$$

Numbers:
$$\sqrt{i^2} = \sqrt{-1} = i \text{ (principal value)}.$$

Sets of numbers:
$$\sqrt{i^2} = \pm i \,.$$

Numbers:
$$+i = \sqrt{-1}, \quad -i = -\sqrt{-1}, \quad \sqrt{-1} = i \text{ (principal values)}.$$

Sets of numbers:
$$\sqrt{-1} = \pm i \,.$$

Numbers: $\sqrt{1} = 1$ (principal value).

Sets of numbers: $\sqrt{1} = \pm 1$.

Note: In the equation $\sqrt{z} = 4$, $\sqrt{z}$ is a *number*, because here $4 \cdot 4 = z$, or $z = 16$.

Note: $\dfrac{-1}{1} = \dfrac{1}{-1}$ is *correct*, $\sqrt{\dfrac{-1}{1}} = \sqrt{\dfrac{1}{-1}}$ is *correct*, but $\dfrac{\sqrt{-1}}{\sqrt{1}} = \dfrac{\sqrt{1}}{\sqrt{-1}}$ is *wrong* because

$\sqrt{\dfrac{-1}{1}} \neq \dfrac{\sqrt{1}}{\sqrt{-1}}$. The rule is *not* $\sqrt{\dfrac{z}{w}} = \dfrac{\sqrt{w}}{\sqrt{z}}$, but $\sqrt{\dfrac{z}{w}} = \dfrac{\sqrt{z}}{\sqrt{w}}$. The theorem is *exactly*

$\sqrt{\dfrac{z_1}{z_2}} = \dfrac{\sqrt{z_1}}{\sqrt{z_2}}$. It is *not* $\sqrt{z_3} = \dfrac{\sqrt{z_1}}{\sqrt{z_2}}$, where $z_3 = z_1 \div z_2$. Example: $\sqrt{-1} \neq \dfrac{\sqrt{1}}{\sqrt{-1}}$.

Note: Axioms and theorems are for operations with numbers, not sets of numbers. When we are calculating we are using numbers, not sets of numbers. Example: $\sqrt{4} + i\sqrt{2} - i\sqrt{2} = 2$, even if $\sqrt{4} = \pm 2$, as a *set* of complex numbers.

Remarks: Many students, and even textbook authors, have problems with the notation for powers and roots, confusing *numbers* with *sets* and vice versa. A mathematician once wrote: "[…] $x^2 + 1 = 0$ only when $x = \pm \mathbf{i}$. It is therefore a matter of choice whether we take $\mathbf{i}$ to be a one or a two valued number." Author's (NKO's) comments: *Numbers are unique.* But, of course, if $\mathbf{i}$ *was replaced* by $-\mathbf{i}$ all theorems and all results would continue to be *equivalently* valid. Another mathematician wrote: "Let us assume that the equation $\sqrt{ab} = \sqrt{a}\sqrt{b}$ is true for all a and b. Let $a = -1$ and $b = -4$. Then we have:

$$\sqrt{(-1)(-4)} = \sqrt{4} = 2.$$
$$\sqrt{-1}\sqrt{-4} = (i)\left(i\sqrt{4}\right) = i^2 \cdot 2 = -2.$$

This example made use of **complex numbers**. It can be shown that with the use of complex numbers either a or b can be negative but not both".

Author's (NKO's) comments: Firstly, the theorems must be used with complete accuracy and precision. Secondly, the *context* (numbers or sets of numbers) must be taken into account. Thirdly, we need the following theorems: $\sqrt{z}\sqrt{z} = \sqrt{z \cdot z} = z$,

$\sqrt{-z}\sqrt{-z} = \sqrt{(-z) \cdot (-z)} = -z$, $\sqrt{z^2} = \pm z$, $\sqrt{z} = \pm w$, where $(\pm w)^2 = z$. Note that the main theorem here is *exactly* $\sqrt{z_1 z_2} = \sqrt{z_1}\sqrt{z_2}$. It is *not* $\sqrt{z_3} = \sqrt{z_1}\sqrt{z_2}$, where $z_3 = z_1 z_2$. By the way: As we remember, $\sqrt{z} \geq 0$, where z is any complex number, is *meaningless* and *wrong*. So the correct calculations are: $\sqrt{-1}\sqrt{-4} = \sqrt{(-1)(-4)} = \sqrt{(i^2)(-4)} = \sqrt{i^2}\sqrt{-4} = i \cdot (2i) = -2$

(the *context* here is *numbers*, so we are using the *principal values* of $\sqrt{-4}$ and $\sqrt{i^2}$), or much shorter: $\sqrt{(-1)(-4)} = \sqrt{(-2)(-2)} = -2$, or $\sqrt{(-1)(-4)} = \sqrt{-1}\sqrt{-4} = i \cdot 2i = -2$ (of course, again using the principal values). In *isolation* (as a *set*) the symbol $\sqrt{4} = \pm 2$ because $\sqrt{4} = \sqrt{2 \cdot 2} = 2$ and $\sqrt{4} = \sqrt{(-2) \cdot (-2)} = -2$. Another example: Generally, in *isolation*, the symbol $\sqrt{2}$ is 2-valued (is the *set* $\left\{ \sqrt{2}, -\sqrt{2} \right\}$ *of complex numbers*), but if the *context* is the *complex number* $z = \sqrt{2} + \left(\sqrt{2}\right)i$, say, then $\sqrt{2}$ is a *number* and so is 1-valued (unique).

Remember that the *complex number* $\sqrt{2}$ is the principal value of the term (symbol) $\sqrt{2}$, that is the *positive* square root of 2.

Roots (as sets) are not unique, so it is clear that complex powers (as sets) are multi-valued. Thus careful treatment of powers (and roots) are needed.

Example (sets): $\left(8^{\frac{1}{3}} \right)^4 = 8^{\frac{4}{3}} \neq 16$ as there are *three* cube roots of 8 in the complex-number system, so $\left(8^{\frac{1}{3}} \right)^4 = 8^{\frac{4}{3}}$ are the *simplest possible* expressions here.

What is "multi-valued", really? To find the nth roots of a complex number $w \neq 0$ we have to solve the equation $z^n = w$. This equation has n distinct solutions (roots). Therefore, as sets, $\sqrt[n]{w} = w^{\frac{1}{n}}$ must be n-valued. More generally (talking about *sets* of numbers), for arbitrary complex powers, **the general meaning of z^w must be multi-valued.**

Examples:

The *symbol (set)* $\sqrt{1+i} = \pm \left[\sqrt{\dfrac{1+\sqrt{2}}{\sqrt{2}}} + \left(\sqrt{\dfrac{\sqrt{2}-1}{\sqrt{2}}} \right) i \right]$, and the *symbol (set)* $\sqrt[3]{-1}$ represents *three* different complex numbers.

The solutions (roots) of the cubic equation $x^3 - x = 0$ are $0, 1,$ and -1. ***Tartaglia's cubic formula*** gives the following solution:

$$x = \frac{1}{\sqrt{3}} \left[\left(\sqrt{-1} \right)^{\frac{1}{3}} + \frac{1}{\left(\sqrt{-1} \right)^{\frac{1}{3}}} \right].$$ However $\sqrt[3]{i} = i^{\frac{1}{3}}$ has the following values

$-i, \dfrac{\sqrt{3}}{2} + \dfrac{1}{2}i, \dfrac{-\sqrt{3}}{2} + \dfrac{1}{2}i$. Substituting these values in turn for $\sqrt[3]{i} = i^{\frac{1}{3}}$ in Tartaglia's solution, we get the solutions $0, 1,$ and -1.

Finding $\sqrt{i}$ *algebraically*:

$i = \left(a + bi \right)^2 = a^2 + 2abi - b^2 = \left(a^2 - b^2 \right) + \left(2ab \right)i$, so $a^2 - b^2 = 1$ and $2ab = 1$, or

$a = b = \pm \dfrac{1}{\sqrt{2}}$, so we have $\sqrt{i} = \pm \left[\dfrac{1}{\sqrt{2}} + \dfrac{1}{\sqrt{2}}i \right]$.

Powers and roots of complex numbers must exist because powers and roots are basically Axioms $00 - 17$ (see above) at work, directly and indirectly (as theorems). Here we do not want to try directly to define general powers and general roots of complex numbers (base, radicand, exponent are complex numbers). Instead we will *discover* the definitions and formulas of complex powers and complex roots by using calculation (using basic theorems and axioms).

Obviously we do have the following structure:

$$\left(a + bi \right)^{c + di}$$
$$= \left(a + bi \right)^c \cdot \left(a + bi \right)^{di}$$
$$= \left(a + bi \right)^c \cdot \left[\left(a + bi \right)^d \right]^i$$
$$= \left(e + fi \right) \cdot \left(g + hi \right)^i$$
$$= \left(e + fi \right) \cdot \left(k + li \right)$$
$$= p + qi$$

But how do we calculate $(a + b\mathbf{i})^c$ (or $(a + b\mathbf{i})^d$) and $(g + h\mathbf{i})^i$?

To do these calculations we need the following theoretical tools:

Firstly we need the equation(s):

$$z = (a,\ b) = a + bi$$
$$= r\cos\theta + (r\sin\theta)i$$
$$= r[\cos\theta + (\sin\theta)i]$$

where θ is the angle (here measured in *radians*) between the positive real axis and the vector (complex number) $(\cos\theta,\ \sin\theta)$. Here $-\pi < \theta \leq \pi$, $r = \sqrt{a^2 + b^2}$ (note: r is *real positive*), $a = r\cos\theta$ and $b = r\sin\theta$ (by Pythagoras' theorem), so $z = r\cos\theta + (r\sin\theta)i$, and the point (complex number) (a, b) has polar coordinates $(r,\ \theta)$. θ is often called the **argument** or **amplitude** of the complex number z, is measured in *radians*, and denoted by arg (z) or amp (z), respectively.

There is only one value of θ in radians that satisfies $-\pi < \theta \leq \pi$. This value is called the **principal argument** of z and is denoted by $\theta = \text{Arg } z$.

Note that if $a \neq 0$ then $\tan\theta = \dfrac{b}{a}$ and $\theta = \arctan\dfrac{b}{a}$. The form $z = r\cos\theta + (r\sin\theta)i$ (or $z = r[\cos\theta + (\sin\theta)i]$) is called the **polar form** or **trigonometric form** of the complex number z. Note: ***Any complex number which is written in polar form is understood to be nonzero. (If $z = 0$, then θ is any angle in $-\pi < \theta \leq \pi$.)*** **Caution**: Using $\theta = \arctan\dfrac{b}{a}$, we must pay attention to the quadrant in which z lies, since $\tan\theta$ has period π, so that **the arguments of z and $\bullet z$ have the same tangent.** We'll often restrict to $-\pi < \theta \leq \pi$ (**principal value** of θ) to obtain a *well defined* value of z (the **principal value** of z).

Example:

Expressing $z = -1 - \mathbf{i}$ in *polar form* using the *princial argument* of z:

$$r = |z| = \sqrt{(-1)^2 + (-1)^2} = \sqrt{2}.$$

$$x = r\cos\theta \rightarrow -1 = \sqrt{2}\cos\theta \rightarrow \cos\theta = \frac{-1}{\sqrt{2}}.$$

$$y = r\sin\theta \rightarrow -1 = \sqrt{2}\sin\theta \rightarrow \sin\theta = \frac{-1}{\sqrt{2}}.$$

$\theta = \dfrac{-3\pi}{4} = 135°$ is the only value of θ that satisfies $-\pi < \theta \le \pi$. Thus we have

$$z = \sqrt{2}\cos\frac{-3\pi}{4} + \left(\sqrt{2}\sin\frac{-3\pi}{4}\right)i.$$

More examples (*principal values* only):

$$1 = \cos 0 + (\sin 0)i.$$

$$-1 = \cos\pi + (\sin\pi)i.$$

$$i = \cos\frac{\pi}{2} + \left(\sin\frac{\pi}{2}\right)i.$$

$$-i = \cos\frac{3\pi}{2} + \left(\sin\frac{3\pi}{2}\right)i.$$

$$1 + i = \sqrt{2}\cos\left(\frac{\pi}{4}\right) + \left[\sqrt{2}\sin\left(\frac{\pi}{4}\right)\right]i$$

$$\sqrt{1} = \cos\pi + (\sin\pi)i.$$

$$e^i = \cos(1) + \left[\sin(1)\right]i.$$

Secondly we must know natural logarithms of real numbers:

Natural logarithms have base $e = 2.718281828$, called **Euler's number**, and was discovered by Jacob Bernoulli (1654 – 1705). Then we have $\ln 1 = 0$, $\ln e = 1$, $e^{\ln e} = e$, and $e^{\ln r} = r$.

Thirdly (and lastly) we need Euler's equation:

Euler's equation is $e^{\theta i} = \cos\theta + (\sin\theta)i$ where e is Euler's number and θ $(-\pi < \theta \le \pi)$ is the angle between the positive real axis and the vector (complex number) $(\cos\theta, \sin\theta)$ and

$r = \sqrt{\sin^2\theta + \cos^2\theta} = 1$. Note: **We need a radian measure here**. Euler's equation was discovered by Roger Cotes in 1714. Various proofs of Euler's equation are possible: Using Taylor series, using calculus, or using ordinary differential equations. APPENDIX 1 is devoted to sketching a proof of Euler's equation using Taylor series.

Of course, many values of θ will fit the above equation:

$e^{\theta i} = \cos\theta + (\sin\theta)i$

$= \cos(\theta + 2k\pi) + (\sin\theta + 2k\pi)i$

$= e^{(\theta + 2k\pi)i}$

where $k = \ldots, -3, -2, -1, 0, 1, 2, 3, \ldots$, so we'll restrict to $-\pi < \theta \leq \pi$ (**principal value** of θ) to obtain a well defined value of z (the **principal value** of z). Remember that there is only one value of θ in radians that satisfies $-\pi < \theta \leq \pi$, and that this value is called the *principal argument* of z and is denoted by $\theta = \operatorname{Arg} z$.

Finally we have $z = r[\cos\theta + (\sin\theta)i] = re^{\theta i}$. The form $z = re^{\theta i}$ is called the **exponential form** or **Euler form** of the complex number z. Note: **Any complex number which is written in polar form or exponential form is understood to be nonzero.** (**If $z = 0$, then θ is any angle in $-\pi < \theta \leq \pi$.**)

Now we can do the calculations:

$(a + bi)^c = \left(re^{\theta i}\right)^c = r^c e^{c\theta i}$

$= r^c \left[\cos(c\theta) + \sin(c\theta)i\right]$

$= \left[r^c \cos(c\theta)\right] + \left[r^c \sin(c\theta)\right]i$

where $r = \sqrt{a^2 + b^2}$ and θ ($-\pi \prec \theta \leq \pi$) is the angle between the positive real axis and the vector (a, b)

and

$$(g + hi)^i = \left(re^{\theta i}\right)^i = r^i e^{\theta i i}$$

$$= r^i e^{-\theta} = e^{-\theta} r^i = e^{-\theta} \left(e^{\ln r}\right)^i = e^{-\theta} e^{(\ln r)i}$$

$$= e^{-\theta} \left\{\cos(\ln r) + \left[\sin(\ln r)\right]i\right\}$$

$$= \left[e^{-\theta} \cos(\ln r)\right] + \left[e^{-\theta}\sin(\ln r)\right]i$$

where $r = \sqrt{g^2 + h^2}$ and θ $(-\pi \prec \theta \leq \pi)$ is the angle between the positive real axis and the vector (g, h).

Now, see below, given *a*, *b*, *c*, *d*, and using the formulas above, we can easily calculate *e*, *f* in

$$(a + bi)^{c + di} = e + fi.$$

If you require the *principal argument* of this result, then you may need to add or subtract 2π the proper of number of times from the argument calculated.

Optional exercises:

Show that $i^i = e^{-\frac{\pi}{2}}$ (principal value).

Show that $1^i = 1$ (principal value).

Note that the result of *raising a complex number to a complex power may be a real number.*

Of course, if $(a + bi)^{c + di} = p + qi$ then $a + bi$ is the $(c + di)$th root of $p + qi$. Note that $(a + bi)^{c + di}$ and the $(c + di)$th root of $p + qi$ are in general not unique.

Finally, see below, given *a*, *b*, *c*, *d*, and using the formulas above, we can easily calculate *e*, *f*, *g*, *h* in

$$\sqrt[c + di]{a + bi} = (a + bi)^{\frac{1}{c + di}} = (a + bi)^{e + fi} = g + hi.$$

If you require the *principal argument* of this result, then you may need to add or subtract 2π the proper of number of times from the argument calculated.

Optional exercises:

Show that $\sqrt[i]{i} = e^{\frac{\pi}{2}}$ (principal value) using the formulas above or Theorem 102 below.

Show that $\sqrt[i]{1} = 1$ (principal value) using the formulas above or Theorem 102 below.

Find a formula for the principal value of $(a+bi)^{c+di}$, remembering that $a+bi = re^{\theta i}$ where $-\pi < \theta \leq \pi$ and $r = \sqrt{a^2 + b^2}$ and that

$$\cos(\theta_1 + \theta_2) = \cos\theta_1 \cos\theta_2 - \sin\theta_1 \sin\theta_2$$
$$\sin(\theta_1 + \theta_2) = \sin\theta_1 \cos\theta_2 + \cos\theta_1 \sin\theta_2.$$

Answer (solution): See Theorem 102 below / proof in APPENDIX 2.

Theorem 102

$$(a+bi)^{c+di} = \left\{r^c e^{-d\theta} \cos(c\theta + d \ln r)\right\} + \left\{r^c e^{-d\theta} \sin(c\theta + d \ln r)\right\}i \text{ where } a+bi = re^{\theta i},$$

$-\pi < \theta \leq \pi$, and $r = \sqrt{a^2 + b^2}$.

If you require the *principal argument* of this result, then you may need to add or subtract 2π the proper of number of times from the argument calculated.

Proof: See APPENDIX 2.

Optional exercises:

Show that $z^0 = 1$, and $z^1 = z$.

Show that $e^{2+i} = e^2 \cos 1 + (e^2 \sin 1)i$ (principal value).

Show that $(-i)^i = e^{-2\pi\left(n - \frac{1}{4}\right)}$ where $n = 0, \pm 1, \pm 2, \pm 3, \dots$.

Show that $e^{\frac{\pi}{2}i} = \pm i$.

Theorem 103 (***de Moivre's Formula***)

If $z = r\cos\theta + (r\sin\theta)i$, $n = 0, 1, 2, 3, 4, 5, \ldots$, and $-\pi < \theta \leq \pi$,

then $z^n = (r\cos\theta + (r\sin\theta)i)^n = [r^n\cos(n\theta)] + [r^n\sin(n\theta)]i$.

If you require the *principal argument* of this result, then you may need to add or subtract 2π the proper of number of times from the argument calculated.

Proof: See APPENDIX 3.

Optional exercise: Show that:

$i^1 = i$.

$i^2 = -1$.

$i^3 = 1$.

$z^0 = 1$.

Theorem 104 (***All complex numbers w have n nth roots***)

If the vector (complex number) $w = a + bi$ makes an angle θ $(-\pi < \theta \leq \pi)$ with the positive direction of the real axis, and $|w| = |a + bi| = r$, then $a = r\cos\theta$ and $b = r\sin\theta$ (by Pythagoras' theorem). So we have $w = a + bi = r\cos\theta + (r\sin\theta)i$. *All complex numbers w have n nth roots.* They are (see proof in APPENDIX 4):

$$\left(\sqrt[n]{w}\right)_k = \left(w^{\frac{1}{n}}\right)_k = z_k = \left[\sqrt[n]{r}\cos\left(\frac{\theta}{n} + k\frac{2\pi}{n}\right)\right] + \left[\sqrt[n]{r}\sin\left(\frac{\theta}{n} + k\frac{2\pi}{n}\right)\right]i$$

where $n = 1, 2, 3, \ldots$ and $k = 0, 1, 2, \ldots, n-1$.

Note: *The value of* $z_k = \left(\sqrt[n]{w}\right)_k$ *obtained by taking the principal value of* θ *and* $k = 0$, *is called* **the principal value of** $z = \sqrt[n]{w}$.

Note that this formula gives $\sqrt{-1} = \pm\mathbf{i}$.

Proof: See APPENDIX 4.

Examples:

If $w = a + b\mathbf{i}$, $r = \sqrt{a^2 + b^2}$, and $-\pi < \theta \le \pi$, then the symbol (set) $\sqrt{w} = w^{\frac{1}{2}}$ has the two values (elements):

$$z_1 = \sqrt{r}\cos\frac{\theta}{2} + \left(\sqrt{r}\sin\frac{\theta}{2}\right)i,$$

$$z_2 = \left[\sqrt{r}\cos\left(\frac{\theta}{2}+\pi\right)\right] + \left[\sqrt{r}\sin\left(\frac{\theta}{2}+\pi\right)\right]i = -z_1.$$

The set $\sqrt{16} = \pm 4$, the set $\sqrt{-16} = \pm 4i$, the set $\sqrt{4i} = \pm\left[\sqrt{2} + \left(\sqrt{2}\right)i\right]$.

Again, remember that in the complex-number system the symbols $\sqrt[n]{w} = w^{\frac{1}{n}}$, where $w \ne 0$, are n-valued, and if $z = \sqrt[n]{w}$, then $z_1{}^n = w$, $z_2{}^n = w$, $z_3{}^n = w$, ... , $z_n{}^n = w$.

Optional exercises:

Calculate the square roots of 4 and the fourth roots of 2.

Answers: ± 2, and $\sqrt[4]{2}, \left(\sqrt[4]{2}\right)i, -\sqrt[4]{2}, \left(-\sqrt[4]{2}\right)i$.

Solve $z^8 = 1$ for z.

Solutions:

$1, \dfrac{1}{\sqrt{2}} + \dfrac{1}{\sqrt{2}}i, i, \dfrac{-1}{\sqrt{2}} + \dfrac{1}{\sqrt{2}}i, -1, \dfrac{-1}{\sqrt{2}} - \dfrac{1}{\sqrt{2}}i, -i, \dfrac{1}{\sqrt{2}} - \dfrac{1}{\sqrt{2}}i$.

Solve $z^{\frac{4}{3}} = -4$ for z.

Solutions:

$z_1 = 2 + 2i, \ z_2 = -2 - 2i, \ z_3 = 2 - 2i, \ z_4 = -2 + 2i$.

As we understand, we do not need to *directly* prove theorems such as

$$\left(z^u\right)^v = z^{uv}, \quad z^u z^v = z^{u+v}, \quad z^u \div z^v = z^{u-v},$$

$$\left(z_1 z_2\right)^w = z_1^w z_2^w, \quad \left(\frac{z_1}{z_2}\right)^w = \frac{z_1^w}{z_2^w}, \quad \text{etc,}$$

where z, z_1, z_2, u, v, and w are complex numbers.

The symbol (set) z^w; an overview:

If $z \neq 0$, then the *symbol (set)* z^w is single-valued if and only if w is an integer (0, ± 1, ± 2, ± 3,

...) If $z \neq 0$, w is a real, rational number, and $w = \dfrac{p}{q}$ is in its lowest term (have no common

factors), then z^w has exactly q distinct values (the q qth roots of z^p). If $z \neq 0$ and w is real and and irrational or has a nonzero imaginary part, then z^w has infinitely many values. Where z^w has distinct values, these values differ by factors of the form $e^{2\pi n w i}$. Note that the symbol (set) z^w, where z and w are non-real complex numbers, has infinitely many values.

Spiral patterns of values for powers and roots:

Only powers of z where $|z| = 1$, lie on the *unit circle*. Conversly, any root of one must also lie on the unit circle. In fact, such roots are spaced evenly around the unit circle, with the first one being the real number 1. Looking at a number of powers of z where $|z| < 1$, at once, the powers of z seem to spiral inward, away from the unit circle, toward the origin. For numbers with modules greater than one ($|z| > 0$), the powers spiral outward away from the unit circle.

Example: If $a \neq 0$ and $b \neq 0$, for general parameter t, $e^{t(a+bi)} = e^{at}\left[\cos(bt) + \sin(bt)i\right]$ parametrizes a spiral. If $a > 0$, the spiral runs away from the origin, counterclockwise if $b > 0$, clockwise if $b < 0$. If $a < 0$ the spiral decays toward the origin. If $a = 0$ the equation parametrizes a circle. If $b = 0$ the curve lies on the positive real axis.

F. Natural logarithms of complex numbers

John Bernoulli discovered *natural logarithms of complex numbers* in the early 1700s.

Let $z \neq 0$, then we know that $z = re^{\theta i} = re^{(\theta + k2\pi)i} = e^{\log r}e^{(\theta + k2\pi)i} = e^{\log r + (\theta + k2\pi)i}$.

So, of course, $\log z = \log r + (\theta + k2\pi)i$, where $-\pi < \theta \leq \pi$, and $k = 0, \pm 1, \pm 2, \pm 3,$
. . . .

Note that **negative real numbers do have a complex logarithm** and that **only positive real numbers have real-valued logarithms**.

The **principal value** of $\log z$ is $\text{Log } z = \log r + \theta i = \log |z| + [\text{Arg } (z)]i$ where $\text{Arg}(z)$ is the unique real θ, called the **principal argument**, which, if $z = a + bi$, satisfies $a = r\cos\theta, \ b = r\sin\theta, \ -\pi < \theta \leq \pi$, and, **if $z = 0$ (but here $z \bullet 0$) then $r = 0$, and arg(z) is any real number**.

Note that the notations $\text{Ln } r$ and $\ln r$, rather than $\text{Log } r$ and $\log r$, are also frequently used.

Example: $\text{Log}(-4i) = \ln 4 + \dfrac{\pi}{2}i$.

Optional exercises:

Use $\text{Log } z$ to prove the following theorems:

($z, z_1, z_2, u, v,$ and w are complex numbers)

$$\left(z^u\right)^v = z^{uv}.$$

$$z_1^w z_2^w = \left(z_1 z_2\right)^w.$$

$$\frac{z_1^w}{z_2^w} = \left(\frac{z_1}{z_2}\right)^w.$$

$$z^u z^v = z^{u+v}.$$

$$z^u \div z^v = z^{u-v}.$$

Remember that the powers above are generally not unique.

G. Functions of a complex variable: An example:

Given $z = x + yi$ (in a z plane), $w = u + vi$ (in a w plane) and $w = f(z) = \dfrac{1}{z}$. Then we have

$u + vi = \dfrac{1}{x + yi}$. Note: $(u, v) \neq (0, 0)$ and $u > 0$ (see below). Given $x = 1$, then

$u + vi = \dfrac{1}{1 + yi} \rightarrow u + vi = \dfrac{1 \cdot (1 - yi)}{(1 + yi)(1 - yi)} \rightarrow u + vi = \dfrac{1 - yi}{1 + y^2}$

$\rightarrow u = \dfrac{1}{1 + y^2}$ and $v = \dfrac{-y}{1 + y^2} \rightarrow \dfrac{v}{u} = -y \rightarrow y = -\dfrac{v}{u} \rightarrow u = \dfrac{1}{1 + \left(-\dfrac{v}{u}\right)^2} = \dfrac{u^2}{u^2 + v^2}$

$\rightarrow u(u^2 + v^2) = u^2 \rightarrow u^2 + v^2 = u \rightarrow u^2 - u + v^2 = 0 \rightarrow u^2 - u + \left(\dfrac{1}{2}\right)^2 + v^2 = 0 + \left(\dfrac{1}{2}\right)^2$

$\rightarrow \left(u - \dfrac{1}{2}\right)^2 + v^2 = \left(\dfrac{1}{2}\right)^2 \rightarrow \left(u - \dfrac{1}{2}\right)^2 + (v - 0)^2 = \left(\dfrac{1}{2}\right)^2$. This is the equation of a circle

in the w plane, centre $\left(\dfrac{1}{2}, 0\right)$, radius $\dfrac{1}{2}$, but excluding the point $(0, 0)$ since $u > 0$. (u, v) is

the *image* in the w plane of the point $(1, y)$ in the z plane, as y increases from $-\infty$ to $+\infty$. Note that as y increases from $-\infty$ to $+\infty$ (u, v) moves one cycle in the clockwise direction.

$f(z) = \dfrac{1}{z}$ *maps* the line $x = 1$ onto a circle in the w plane. (See more about complex functions in APPENDIX 7.)

H. More optional exercises (with answers):

Note: There are free ***complex number calculators*** on the Internet.

1. Using only axioms, give a proof for the following:

    $$\dfrac{1}{z_1} \cdot \dfrac{1}{z_2} = \dfrac{1}{z_1 z_2}$$

    and

    $$\dfrac{1}{z_1} + \dfrac{1}{z_2} = \dfrac{z_1 + z_2}{z_1 z_2}.$$

2. Using only axioms, prove that if the product of two complex numbers is zero, then at least one factor must be zero.

3. Show that $5 = 5 + 0i$.

4. Show that $3 + \sqrt{-4} = 3 + 2i$.

5. Show that $2 + \sqrt{-5} = 2 + \left(\sqrt{5}\right)i$.

6. Show that $\sqrt{5} + \sqrt{-5} = \sqrt{5} + \left(\sqrt{5}\right)i$.

7. Show that $3 - 2\sqrt{-1} = 3 - 2i$.

8. Show that $(2 + 3i) + (4 + i) = 6 + 4i$.

9. Show that $(5 + 7i) - (3 - 2i) = 2 + 9i$.

10. Show that $(\sqrt{2} - i) - i(1 - \sqrt{2}\,i) = -2i$.

11. Show that $(1 + i)(1 - 2i) = -\dfrac{1}{5} + \dfrac{3}{5}i$.

12. Show that $\dfrac{1}{2}(1 + i)(1 + i^{-8}) = 1 + i$.

13. Show that $\dfrac{1}{i} = -i$.

14. Use the result $\dfrac{1}{i} = -i$ to readily show that $\dfrac{a + bi}{i} = b - ai$.

15. Show that $\dfrac{1}{i + 1} = \dfrac{1 - i}{2}$.

16. Show that $\dfrac{1}{i} + \dfrac{3}{1 + i} = \dfrac{3}{2} - \dfrac{5}{2}i$.

17. Show that $\dfrac{2 + 3i}{4 + i} = \dfrac{11}{17} + \dfrac{10}{17}i$.

18. Show that $\dfrac{3 + 2i}{4 - 3i} = \dfrac{6}{25} + \dfrac{17}{25}i$.

19. Let $\dfrac{x - yi}{x + yi} = a + bi$. Show that $a^2 + b^2 = 1$.

20. Show that $i^3 = 0 - i$.

21. Show that $i^4 = 1 + 0i$.

22. Show that $i^{217} = i$.

23. Show that $i^{602} = -1$.

24. Show that $i^{1803} = -i$.

25. Show that $(1 + i)^4 = -4$.

26. Show that the term (set) $\sqrt[4]{-4} = \{1+i, 1-i, -1+i, -1-i\}$.

27. Show that if $z = a + b\mathbf{i}$, then

$$z^2 = (a^2 - b^2) + (2ab)i,$$

$$z^3 = (a^3 - 3ab^2) + (3a^2b - b^3)i.$$

28. Show that $2 + 2i = 2\sqrt{2}\cos\dfrac{\pi}{4} + \left(2\sqrt{2}\sin\dfrac{\pi}{4}\right)i$.

29. Show that $1 - i = \sqrt{2}\cos\left(-\dfrac{\pi}{4}\right) + \left[\sqrt{2}\sin\left(-\dfrac{\pi}{4}\right)\right]i$.

30. Show that $-1 + i = \sqrt{2}\cos\dfrac{3\pi}{4} + \left(\sqrt{2}\sin\dfrac{3\pi}{4}\right)i$.

31. Show that $(-\mathbf{i})^{-1} = \mathbf{i}$.

32. Show that $\dfrac{\sqrt{2}}{2} + \dfrac{\sqrt{2}}{2}\mathbf{i}$ and $-\dfrac{\sqrt{2}}{2} - \dfrac{\sqrt{2}}{2}\mathbf{i}$ are square roots of $\mathbf{i}$.

33. Show that the symbol (set) $\sqrt{1+i} = \pm\left(\sqrt{\dfrac{1+\sqrt{2}}{2}} + \sqrt{\dfrac{\sqrt{2}-1}{2}}i\right)$.

34. Show that the symbol (set) $\sqrt{2i} = \pm(1+i)$.

35. Show that $2^i = 0.7692389\ldots + 0.6389612\ldots\mathbf{i}$.

36. Show that $16^{-i} = -0.932687\ldots - 0.360686\ldots\mathbf{i}$.

37. Use the internet to learn more about Euler's identity and Euler's number:

     Euler's identity: $e^{\pi i} + 1 = 0$,

     Euler's number: $e = 2.718281828459\ldots$,

     Euler's Identity is called The Greatest Equation Ever.

38. Show that $\mathbf{i}^i = 0.207879576\ldots$ (principal value).

          Certainly, *there is nothing imaginary about imaginary numbers.*

39. Given:

     $$i = e^{\frac{\pi}{2}i},$$

     $i =$ the imaginary unit,

     $e =$ Euler's number $= 2.718281828459\ldots$.

     Calculate $\sqrt[i]{i}$ and i^i (principal values).

     Answers (solutions):

     $$\sqrt[i]{i} = i^{\frac{1}{i}} = \left(e^{\frac{\pi}{2}i}\right)^{\frac{1}{i}} = e^{\frac{\pi}{2}} \text{ (principal value)},$$

$$i^i = \left(e^{\frac{\pi}{2}i}\right)^i = e^{-\frac{\pi}{2}} \quad \text{(principal value)}.$$

40. Show that $\sqrt[i]{-1} = 23.14069...$ (principal value).
41. Show that $1^i = 1$ and $1^z = 1$ (principal values).
42. Show that $|-3 + 2i| = \sqrt{13}$.
43. Show that $|1 + 4i| = \sqrt{17}$.
44. Show that $|zw| = |z| \cdot |w|$.

45. Show that $\left|\dfrac{z}{w}\right| = \dfrac{|z|}{|w|}$.

46. Show that for any complex number z, $z\overline{z} = |z|$.

47. Show that $\overline{\overline{z}} = z$.
48. Show that $z = \overline{z}$ if and only if z is a real number.
49. Show that $|-z| = |z|$ and $\|z\| = |z|$.

50. Show that $|\overline{z}| = |z|$.

51. Show that $|z| = 0$ if and only if $z = 0$.

52. Show that $|z_1 + z_2|^2 = z_1\overline{z_1} + z_2\overline{z_2} + z_2\overline{z_1} + z_1\overline{z_2}$.

53. Show that $\operatorname{Re} z = \dfrac{z + \overline{z}}{2}$.

54. Show that $\operatorname{Im} z = \dfrac{z - \overline{z}}{2i}$.

55. Show that $\operatorname{Re}(iz) = -\operatorname{Im}(z)$ for any complex number z.
56. Show that $\operatorname{Im}(iz) = \operatorname{Re}(z)$ for any complex number z.
57. Show that if z is a complex number such that $|z| = \operatorname{Re} z$, then z is a nonnegative real number.
58. Is it true that $\operatorname{Re}(zw) = (\operatorname{Re} z)(\operatorname{Re} w)$? Answer: No. Counterexample: $z = w = i$.
59. Show that $z + \overline{z}$ and $z\overline{z}$ are real.
60. Show that if z and w are complex numbers such that $z + w$ is a real number and zw is a negative real number, then z and w are both real numbers.
61. Show that for any complex number z the imaginary part of the number $z + \overline{z}$ is zero.
62. Show that z is *pure imaginary* if and only if $\overline{z} = -z$.
63. Show that if $(\overline{z})^2 = z^2$, then z is either real or pure imaginary.
64. Show that $\overline{z^n} = (\overline{z})^n$.

75

65. Show that if $|z| = 1$, then $|z^n| = 1$.

66. Show that if $z^n = 1$, then $|z| = 1$.

67. Show that Im $z = 0$ if and only if $z = \overline{z}$.

68. Show that $0 = \overline{0}$.

69. Show that every real number is its own conjugate.

70. Show that the conjugate of each pure imaginary number is its negative.

71. Show that if $z_2 \neq 0$, then $\dfrac{z_1}{z_2} = \dfrac{1}{|z_2|^2} z_1 \overline{z}_2$.

72. Show that if $z_2 \neq 0$, then $\dfrac{z_1}{z_2} = \dfrac{z_1 \overline{z}_2}{z_2 \overline{z}_2}$.

73. Using the formula $\dfrac{z_1}{z_2} = \dfrac{1}{|z_2|^2} z_1 \overline{z}_2$, show that $\dfrac{1}{(3+4i)^2} = \dfrac{-7-24i}{625}$.

74. Using the formula $\dfrac{z_1}{z_2} = \dfrac{z_1 \overline{z}_2}{z_2 \overline{z}_2}$, show that $\dfrac{i}{1+i} = -\dfrac{1}{2} + \dfrac{1}{2} i$.

75. Show that $z^{-1} = \dfrac{\overline{z}}{|z|^2}$ $(z \neq 0)$.

76. Show that, if $z = re^{\theta i}$, then $\overline{z} = re^{-\theta i}$.

77. If b is real, show that $|a^b| = |a|^b$.

78. Show that $|e^{a+bi}| = e^a$.

79. Show that for any complex number z if $z^2 = 0$ then $z = 0$.

80. Show that $e^{z+w} = e^z e^w$, where z and w are complex numbers.

81. Show that $e^{\frac{\pi}{2}i} = i$.

82. Show that $e^{\pi i} = -1$.

83. Shoe that $e^{\frac{3\pi}{2}i} = -i$.

84. Show that $e^{2\pi i} = 1$.

85. Show that the complex number $z = i^i = e^{-\frac{\pi}{2}}$.

86. Show that the term (set) $(-i)^i = e^{-2\pi(n-\frac{1}{4})}$, where n is an integer.

87. Show that the term (set) $i^{2i} = e^{-(4n+1)\pi}$, where n is an integer.

88. Show that the term (set) $(1+i)^i = e^{\frac{-(8n+1)\pi}{4} + \frac{(\log_e 2)}{2} i}$, where n is an integer.

89. Prove directly that $\sqrt[-w]{z} = \dfrac{1}{\sqrt[w]{z}}$, where $z \neq 0$ and $w \neq 0$.

90. Show that if $3i(2 + 5i) = z + 6i$, then $z = -15$.

91. Show that if $z^2 = 3 - 4i$, then $z = \pm(2 - i)$.

92. Show that if $z^2 - 2z + 2 = 0$, then $z = 1 \pm i$.

93. Show that if $3z^2 + 2z + 1 = 0$, then $z = -\dfrac{1}{3} \pm \dfrac{\sqrt{2}}{3}i$.

94. Solve the equation: $z^4 + i = 0$.

    Answer:
    $$z = \cos\left(\frac{3\pi}{8} + k\frac{2\pi}{4}\right) + \left[\sin\left(\frac{3\pi}{8} + k\frac{2\pi}{4}\right)\right]i,$$

    $k = 0, 1, 2, 3, 4, 5, \ldots$.

95. Solve the equation: $z^5 - 2 = 0$.

    Answer:
    $$z_k = \sqrt[5]{2}\cos\left(k\frac{2\pi}{5}\right) + \left(\sqrt[5]{2}\sin k\frac{2\pi}{5}\right)i,$$

    $k = 0, 1, 2, 3, 4$.

96. Is the *symbol* (*set*) $\sqrt{1 + \sqrt{i}}$ 2-valued or 4-valued?

    Answer. 4-valued.

97. Is the *complex number* $\sqrt{1 + \sqrt{i}}$ 4-valued or 1-valued (unique)?

    Answer. Unique (= the *principal value* (*principal root*) of $\sqrt{1 + \sqrt{i}}$).

98. Solve the equation: $z^2 + 2z + (1 - i) = 0$.

    Answer:
    $$z_1 = \left(-1 + \frac{1}{\sqrt{2}}\right) + \left(\frac{1}{\sqrt{2}}\right)i,$$

    $$z_2 = \left(-1 - \frac{1}{\sqrt{2}}\right) - \left(\frac{1}{\sqrt{2}}\right)i.$$

99. If $z = (a, b)$ and $z^2 + z + 1 = 0$, then $(a, b)(a, b) + (a, b) + (1, 0) = (0, 0)$. Show, by

    solving a pair of simultaneous equations in a and b, that $z_1 = (a_1, b_1) = -\dfrac{1}{2} + \dfrac{\sqrt{3}}{2}i$ and

    $$z_2 = (a_2, b_2) = -\frac{1}{2} - \frac{\sqrt{3}}{2}i.$$

100. Show that $z = \sqrt{a+bi} = \sqrt{\dfrac{r+a}{2} + \left(\dfrac{b}{\sqrt{2(r+a)}}\right)i}$, where $z \neq 0$, and

$r = |a+bi| = \sqrt{a^2 + b^2}$. Note that *the real part of the principal value of the set of numbers* $\sqrt{a+bi}$ *is always non-negative, that is the real part of the complex number* $z = \sqrt{a+bi}$ *is always non-negative.*

101. Show that the sum of the n distinct nth roots of 1 is zero.

102. Show directly that $(z + w)(z + w) = z^2 + 2zw + w^2$.

103. Show directly that $(z - w)(z - w) = z^2 - 2zw + w^2$.

104. Show directly that $(z + w)(z - w) = z^2 - w^2$.

105. Show directly that $z^2 - 1 = (z + 1)(z - 1)$.

106. Show directly that $z^2 - i^2 = (z + i)(z - i)$.

107. Show that for all real numbers a, b, c, d we have

$$\left(a^2 + b^2\right)\left(c^2 + d^2\right) = \left(ac - bd\right)^2 + \left(ad + bc\right)^2.$$ Hint: Apply the rule $|z||w| = |zw|$ to

$z = a + bi$ and $w = c + di$. Here complex numbers serve to *discover* the Two-Square Theorem.

Sketching a proof of Euler's equation (using Taylor series)

First we need *Euler's number:* $\lim_{n \to \infty} \left(1 + \dfrac{1}{n} \right)^n = e$.

Next we need to know this definition: $n! = 1 \cdot 2 \cdot 3 \cdot ... \cdot (n-1) \cdot n$. Note that $0! = 1$ and $1! = 1$.

And we need this theorem: $e^x = \dfrac{x^0}{0!} + \dfrac{x^1}{1!} + \dfrac{x^2}{2!} + \dfrac{x^3}{3!} + ... + \dfrac{x^n}{n!} + ...$ where x is a real number.

From the above theorem follows this theorem: $e^z = \dfrac{z^0}{0!} + \dfrac{z^1}{1!} + \dfrac{z^2}{2!} + \dfrac{z^3}{3!} + ... + \dfrac{z^n}{n!} + ...$ and, of

course, this theorem: $e^{\theta i} = \dfrac{\theta^0 i^0}{0!} + \dfrac{\theta^1 i^1}{1!} + \dfrac{\theta^2 i^2}{2!} + \dfrac{\theta^3 i^3}{3!} + ... + \dfrac{\theta^n i^n}{n!} + ...$. Now we need these

theorems:

$$\cos\theta = 1 - \dfrac{\theta^2}{2!} + \dfrac{\theta^4}{4!} - ... ,$$

$$\sin\theta = \theta - \dfrac{\theta^3}{3!} + \dfrac{\theta^5}{5!} - ...$$

where θ is a real number.

So, finally:

$$e^{\theta i} = \dfrac{\theta^0 i^0}{0!} + \dfrac{\theta^1 i^1}{1!} + \dfrac{\theta^2 i^2}{2!} + \dfrac{\theta^3 i^3}{3!} + ... + \dfrac{\theta^n i^n}{n!} + ...$$

$$= \left(1 - \dfrac{\theta^2}{2!} + \dfrac{\theta^4}{4!} - ... \right) + \left(\theta - \dfrac{\theta^3}{3!} + \dfrac{\theta^5}{5!} - ... \right) i$$

$$= \cos\theta + (\sin\theta) i.$$

By the way: If z and w are complex numbers, then we have $e^{zi} = \cos z + (\sin z) i$, as
expected.

Proof of formula: Complex exponents

Remember: *Any complex number which is written in exponential form or polar form is understood to be nonzero. (If $z = 0$, then θ is **any** angle in $-\pi < \theta \leq \pi$.)*

We want to find a formula for the principal value of $(a+bi)^{c+di}$, remembering that

$a+bi = re^{\theta i}$, where $-\pi < \theta \leq \pi$, $r = \sqrt{a^2+b^2}$ and
$\cos(\theta_1 + \theta_2) = \cos\theta_1 \cos\theta_2 - \sin\theta_1 \sin\theta_2$, $\sin(\theta_1 + \theta_2) = \sin\theta_1 \cos\theta_2 + \cos\theta_1 \sin\theta_2$.

First note that $r = e^{\ln r}$ and $r^{di} = \left(e^{\ln r}\right)^{di} = e^{(d\ln r)i} = \cos(d\ln r) + \left[\sin(d\ln r)\right]i$.

Let $z = a + bi$ and $w = c + di$. Then:

$$z^w = (a+bi)^{c+di} = \left(re^{\theta i}\right)^{c+di} = \left(re^{\theta i}\right)^c \left(re^{\theta i}\right)^{di} = r^c e^{c\theta i} r^{di} e^{d\theta i}$$

$$= r^c e^{-d\theta} r^{di} e^{c\theta i} = r^c e^{-d\theta} e^{c\theta i} e^{(d\ln r)i}$$

$$= \left(r^c e^{-d\theta}\right)\left\{\cos(c\theta) + \left[\sin(c\theta)\right]i\right\} + \left\{\cos(d\ln r) + \left[\sin(d\ln r)\right]i\right\}$$

$$= \left(r^c e^{-d\theta}\right)\left\{\left[\cos(c\theta)\cdot\cos(d\ln r) - \sin(c\theta)\cdot\sin(d\ln r)\right] + \left[\cos(c\theta)\cdot\sin(d\ln r) + \sin(c\theta)\cdot\cos(d\ln r)\right]i\right\}$$

$$= \left(r^c e^{-d\theta}\right)\left[\cos(c\theta + d\ln r) + \sin(c\theta + d\ln r)i\right]$$

$$= \left[r^c e^{-d\theta}\cos(c\theta + d\ln r)\right] + \left[r^c e^{-d\theta}\sin(c\theta + d\ln r)\right]i.$$

If you require the *principal argument* of this result, then you may need to add or subtract 2π the proper of number of times from the argument calculated.

APPENDIX 3

Proof of de Moivre's formula (by induction)

Let $z \neq 0$, then

$$z = z^1 = r^1 e^{(1\theta)i}$$
$$= r^1 \cos(1\theta) + \left[r^1 \sin(1\theta) \right] i$$

$$z^2 = zz^1 = r^1 e^{(1\theta)i} r^1 e^{(1\theta)i} = r^{1+1} e^{1\theta+1\theta} = r^2 e^{2\theta}$$
$$= r^2 \cos(2\theta) + r^2 \sin(2\theta)i$$

$$z^3 = zz^2 = r^1 e^{(1\theta)i} r^1 e^{(1\theta)i} r^1 e^{(1\theta)i} = r^{1+1+1} e^{1\theta+1\theta+1\theta} = r^3 e^{3\theta}$$
$$= r^3 \cos(3\theta) + r^3 \sin(3\theta)i$$

....

....

....

$$z^n = zz^{n-1} = r^1 e^{(1\theta)i} r^1 e^{(1\theta)i} r^1 e^{(1\theta)i} \dots r^1 e^{(1\theta)i}$$
$$= r^{1+1+1+\dots+1} e^{1\theta+1\theta+1\theta+\dots+1\theta} = r^n e^{n\theta}$$
$$= r^n \cos(n\theta) + r^n \sin(n\theta)i.$$

Note that formally, $z^0 = r^0 \cos(0 \cdot \theta) + \left[r^0 \sin(0 \cdot \theta) \right] i = 1 \cdot 1 + \left[1 \cdot 0 \right] i = 1$. So the formula is valid for $n = 0, 1, 2, 3, \dots$.

If you require the *principal argument* of this result, then you may need to add or subtract 2π the proper of number of times from the argument calculated.

By the way: de Moivre's theorem is true in a more general setting than above. If z and w are complex numbers, then $\left[\cos z + (\sin z)i \right]^w = \cos(wz) + \left[\sin(wz) \right] i$. But now $\left[\cos z + (\sin z)i \right]^w$ is a *multivalued function* while $\cos(wz) + \left[\sin(wz) \right] i$ is *not*. Therefore $\cos(wz) + \left[\sin(wz) \right] i$ is one value of $\left[\cos z + (\sin z)i \right]^w$.

81

Proof of formula for the *n* *n*th roots

To find the *n*th root(s) of a complex number $w \neq 0$ we have to solve the equation $z^n = w$. Let $w = r\cos\varphi + (r\sin\varphi)i$ and $z = \rho\cos\theta + (\rho\sin\theta)i$, then, by *de Moivre's formula*:

$$\left(\rho\cos\theta + (\rho\sin\theta)i\right)^n$$
$$= \left[\rho^n \cos(n\theta)\right] + \left[\rho^n \sin(n\theta)\right]i$$
$$= r\cos\varphi + (r\sin\varphi)i.$$

The equation above is fulfilled if $\rho^n = r$ and $n\theta = \varphi$, and we have

$$z = \left[\sqrt[n]{r}\cos\left(\frac{\varphi}{n}\right)\right] + \left[\sqrt[n]{r}\sin\left(\frac{\varphi}{n}\right)\right]i \text{ where } \sqrt[n]{r} \text{ is the positive } n\text{th root of the positive}$$

number *r*. But this is not the only solution of the equation $z^n = w$. Of course, it is also fulfilled for all angles satisfying $\theta = \dfrac{\varphi + k2\pi}{n} = \dfrac{\varphi}{n} + \dfrac{k2\pi}{n}$, where $k = 0, \pm 1, \pm 2, \pm 3, \ldots$. But only the values $k = 0, 1, 2, 3, \ldots (n-1)$ give different values of z. So, the complete solution of the equation $z^n = w$ is given by

$$z = \left[\sqrt[n]{r}\cos\left(\frac{\varphi}{n} + \frac{k2\pi}{n}\right)\right] + \left[\sqrt[n]{r}\sin\left(\frac{\varphi}{n} + \frac{k2\pi}{n}\right)\right]i, \text{ where } k = 0, 1, 2, 3, \ldots, (n-1). \text{ Every}$$

non-zero complex number has exactly *n* distinct *n*th roots. That is: Every non-zero complex number has two square roots, three cube roots, four fourth roots, and so on. The roots have the same *r* and the θs differ by $\dfrac{k2\pi}{n}$. In the complex plane the *n* *n*th roots are the *n* vertices of a regular polygon with *n* sides. The *n* *n*th roots are equally spaced on the circle of radius $\sqrt[n]{r}$. The first root has $\theta = \dfrac{\varphi}{n}$, the second root has $\theta = \dfrac{\varphi}{n} + \dfrac{2\pi}{n}$, and so on. Note: The value of $z_k = \left(\sqrt[n]{w}\right)_k$ obtained by taking the principal value of φ and $k = 0$, is called the

principal value of $z = \sqrt[n]{w}$.

More on exponential representation (exponential form)

Remember: *Any complex number which is written in exponential form or polar form is understood to be nonzero. (If $z = 0$, then θ is **any** angle in $-\pi < \theta \leq \pi$.)*

Let r and θ be polar coordinates of the point (complex number) $(a, b) \neq (0, 0)$, then (as we know):

$e^{\theta i} = \cos\theta + (\sin\theta)i$ or

$\cos\theta + (\sin\theta)i = e^{\theta i}$

$z = a + bi$

$a = r\cos\theta$

$b = r\sin\theta$

$-\pi < \theta \leq \pi$

$z = a + bi = r\cos\theta + (r\sin\theta)i$

$= r\left[\cos\theta + (\sin\theta)i\right] = re^{\theta i}$

(θ is measured in *radians*).

Remember that $re^{\theta i} = r\cos\theta + (r\sin\theta)i$, for any θ, see APPENDIX 1.

And remember that

$r\cos\theta + (r\sin\theta)i = \left[r\cos(\theta + n2\pi)\right] + \left[r\sin(\theta + n2\pi)\right]i$

$n = 0, \pm1, \pm2, \pm3, \dots$.

So, $z = re^{\theta i}$ is only one of an infinite number of possibilities for the exponential form of z:

$z = re^{\theta i} = re^{(\theta + n2\pi)i}$.

Multiplicaton:

Let $z_1 \neq 0$, $z_2 \neq 0$, then

$$z_3 = z_1 z_2 = \left(r_1 e^{\theta_1 i}\right)\left(r_2 e^{\theta_2 i}\right) = r_1 r_2 \left(e^{\theta_1 i} e^{\theta_2 i}\right) = r_1 r_2 e^{(\theta_1 i + \theta_2 i)} = r_1 r_2 e^{(\theta_1 + \theta_2)i}$$

$$= \left[(r_1 r_2)\cos(\theta_1 + \theta_2)\right] + \left[(r_1 r_2)\sin(\theta_1 + \theta_2)\right]i = r_3 \cos\theta_3 + (r_3 \sin\theta_3)i$$

$$= r_3 e^{\theta_3 i}.$$

If you require the *principal argument* of this result, then you may need to add or subtract 2π the proper of number of times from the argument calculated.

Obviously, ***multiplication of complex numbers is much easier to perform in the exponential form than in the standard form (Cartesian form). Addition and subtraction of complex numbers are easier to perform in standard form than in exponential form.***

Multiplication of complex numbers amounts to multiplying their moduli and adding their arguments.

Geometrically, addition and subtraction of complex numbers are just the usual addition of vectors by the parallelogram law (remembering $z_1 - z_2 = z_1 + (-z_2)$).

Optional exercise: Draw an illustration of multiplication of complex numbers.

Division:

Let $z_1, z_2 \neq 0$, then

$$z_3 = \frac{z_1}{z_2} = \frac{r_1 e^{\theta_1 i}}{r_2 e^{\theta_2 i}} = \frac{r_1}{r_2} \cdot \frac{e^{\theta_1 i}}{e^{\theta_2 i}} = \frac{r_1}{r_2} e^{\theta_1 i - \theta_2 i} = \frac{r_1}{r_2} e^{(\theta_1 - \theta_2)i} = \frac{r_1}{r_2}\left\{\cos(\theta_1 - \theta_2) + \left[\sin(\theta_1 - \theta_2)\right]i\right\}$$

$$= \left[\frac{r_1}{r_2}\cos(\theta_1 - \theta_2)\right] + \left[\frac{r_1}{r_2}\sin(\theta_1 - \theta_2)\right]i = r_3 \cos\theta_3 + (r_3 \sin\theta_3)i$$

$$r_3\left[\cos\theta_3 + (\sin\theta_3)i\right] = r_3 e^{\theta_3 i}.$$

If you require the *principal argument* of this result, then you may need to add or subtract 2π the proper of number of times from the argument calculated.

By the way: The ***multiplicative inverse*** of $z = re^{\theta i}$ is

$$z^{-1} = \frac{1}{z} = \frac{1 \cdot e^{0i}}{r \cdot e^{\theta i}} = \frac{1}{r} \cdot \frac{e^{0i}}{e^{\theta i}} = \frac{1}{r} \cdot e^{0i - \theta i} = \frac{1}{r} e^{-\theta i}.$$

Obviously, *division of complex numbers is much easier to perform in the exponential form than in the standard form* (*Cartesian form*). (*Addition and subtraction of complex numbers are easier to perform in standard form than in exponential form.*)

Division of complex numbers amounts to dividing their moduli and subtracting their arguments.

Optional exercise: Draw an illustration of division of complex numbers.

Powers:

$$z^n = \left(re^{\theta i}\right)^n = r^n \left(e^{\theta i}\right)^n = r^n e^{(n\theta)i}$$

$$n = 0, \pm 1, \pm 2, \pm 3, \dots.$$

If you require the *principal argument* of this result, then you may need to add or subtract 2π the proper of number of times from the argument calculated.

Obviously, *exponentiation of complex numbers is much easier to perform in the exponential form than in the standard form* (*Cartesian form*).(*Addition and subtraction of complex numbers are easier to perform in standard form than in exponential form.*)

Proving the theorem $z_1^n z_2^n = \left(z_1 z_2\right)^n$ for *principal arguments* and *principal values*, by using polar coordinates: Let $z_1 \neq 0$, $z_2 \neq 0$, n an integer, then

$$z^n = \left(re^{\theta i}\right)^n = r^n \left(e^{\theta i}\right)^n = r^n e^{(n\theta)i}$$

$$n = 0, \pm 1, \pm 2, \pm 3, \dots.$$

$$z_1^n z_2^n = \left(r_1 e^{\theta_1 i}\right)^n \left(r_2 e^{\theta_2 i}\right)^n = r_1^n e^{n\theta_1 i} r_1^n e^{n\theta_2 i} = r_1^n r_1^n e^{n\theta_1 i} e^{n\theta_2 i} = \left(r_1 r_2\right)^n \left(e^{(\theta_1 + \theta_2)i}\right)^n$$

$$= \left[\left(r_1 r_2\right)\left(e^{(\theta_1 + \theta_2)i}\right)\right]^n = \left[\left(r_1 e^{\theta_1 i}\right)\left(r_2 e^{\theta_2 i}\right)\right]^n = \left(z_1 z_2\right)^n.$$

Roots:

Given $w = re^{(\theta + k2\pi)i}$ where $-\pi < \theta \le \pi$
and $z^n = w$, then

$$z = w^{\frac{1}{n}} = \sqrt[n]{re^{(\theta + k2\pi)i}} = \sqrt[n]{r}\sqrt[n]{e^{(\theta + k2\pi)i}} = \sqrt[n]{r}\left(e^{(\theta + k2\pi)i}\right)^{\frac{1}{n}}$$

$$= \sqrt[n]{r}e^{\left(\frac{\theta}{n} + k\frac{2\pi}{n}\right)i}, \text{ where}$$

$k = 0, 1, 2, 3, \ldots, (n-1)$,

$\sqrt[n]{r}$ is the positive nth root of r,

$w^{\frac{1}{n}} = \sqrt[n]{w}$ is the set of the n nth roots of w.
If $-\pi < \theta \le \pi$ and $k = 0$, then z is
the principal root of w, and so is a *number*.

Obviously, ***the extraction of roots of complex numbers is much easier to perform in the exponential form than in the standard form (Cartesian form). (Addition and subtraction of complex numbers are easier to perform in standard form than in exponential form.)***

The nth roots of 1:

Remember that
$z^n = r^n e^{(n\theta)i}$, where
$n = 0, \pm 1, \pm 2, \pm 3, \ldots$.

Then:
$z^n = 1$ $\quad (0^n = 0, \text{ so } z \ne 0, n = 2, 3, 4, \ldots)$
$\left(re^{\theta i}\right)^n = 1$
$r^n e^{(n\theta)i} = 1$
$r^n e^{(n\theta)i} = 1 \cdot e^{0i}$.

Two nonzero complex numbers $z_1 = r_1 e^{\theta_1 i}$ and $z_2 = r_2 e^{\theta_2 i}$ are equal if and only if $r_1 = r_2$ and $\theta_1 = \theta_2 + k2\pi$, where $k = 0, \pm 1, \pm 2, \pm 3, \ldots$.

So,

$$r^n e^{(n\theta)i} = 1$$

$$r^n e^{(n\theta + k2\pi)i} = 1 \cdot e^{0i} \quad (k = 0, \pm 1, \pm 2, \pm 3, \ldots)$$

$$r^n = 1$$

$$r = 1$$

$$n\theta + k2\pi = 0$$

$$\theta = \frac{k2\pi}{n}$$

$$z = 1 e^{\frac{k2\pi}{n}i} = e^{\frac{k2\pi}{n}i}.$$

All z_k lie on the **unit circle** centered at the origin and are equally spaced around that circle every $\dfrac{2\pi}{n}$ radians. Evidently, all the *distinct* nth roots of 1 are $z_k = e^{\frac{k2\pi}{n}i}$ where $k = 0, 1, 2, 3,$ ... $(n - 1)$. So the number of nth roots of 1 is n. When $n = 2$, $z = \pm 1$. When $n \geq 3$ the n nth roots lie at the n vertices of a *regular polygon* of n sides. The polygons are, of course, inscribed in the unit circle centered at the origin, and it has one vertex at the point corresponding to the root $z = 1$ ($k = 0$).

Proving the theorem $\sqrt[n]{z_1}\sqrt[n]{z_2} = \sqrt[n]{z_1 z_2}$ for *principal arguments* and *principal values*, by using polar coordinates: Let $z_1 \neq 0$, $z_2 \neq 0$, n an integer, then

$$\sqrt[n]{z_1}\sqrt[n]{z_2} = \left(\sqrt[n]{r_1}\, e^{\frac{\theta_1 i}{n}}\right)\left(\sqrt[n]{r_2}\, e^{\frac{\theta_2 i}{n}}\right) = \sqrt[n]{r_1 r_2}\, e^{\frac{(\theta_1 + \theta_2)i}{n}} = \sqrt[n]{z_1 z_2}.$$

APPENDIX 6

Translation, rotation, expansion, contraction, reflexion, inversion, straight line, circle, and ellipse in the complex plane

Translation:

Example: $w = z + 5$ is a *translation* of each point z to a position 5 units to the right.

Rotation:

Let

$$z_3 = z_1 z_2 = r_1 e^{\theta_1 i} r_2 e^{\theta_2 i} = r_1 r_2 e^{\theta_1 i} e^{\theta_2 i} = r_1 r_2 e^{(\theta_1 i + \theta_2 i)} = r_1 r_2 e^{(\theta_1 + \theta_2)i} =$$
$$\left[r_1 r_2 \cos(\theta_1 + \theta_2) \right] + \left[r_1 r_2 \sin(\theta_1 + \theta_2) \right] i.$$

So, the effect of *multiplication* by z_2 is to rotate the complex number z_1 an angle equal to θ_2 and to change its length by the factor r_2.

And, obviously, the effect of *division* by z_2 is to rotate the complex number z_1 an angle equal to $-\theta_2$ and to change its length by the factor $\dfrac{1}{r_2}$.

Example: $w = iz$ rotates each non-zero point z in the positive direction through a *right angle*.

Expansion:

$w = az$, where $a > 1$, is an *expansion*.

Contraction:

$w = az$, where $0 < a < 1$, is a *contraction*.

Reflection:

$w = \overline{z}$ transforms each point z into its *reflection* in the real axis.

Inversion:

$w = \dfrac{1}{z}$ maps every straight line or circle onto a circle or straight line. (Proof omitted.)

Straight line:

A parametric equation of a *straight line* in the complex pane passing through the point *a* in the direction of *b* is $z = a + bt$, where *a* and *b* are complex numbers, and *t* is a real parameter.

Circle:

$|z - a| = r$ is an equation of a *circle* in the complex plane with radius *r* and center at *a*. Note: $|z - a|$ is *length*.

Ellipse:

$|z - d| + |z + d| = 2a$ is an equation of an *ellipse* in the complex plane with foci at $\pm d$ and the *semimajor axis* equals *a*. Note: $|z - d|$ and $|z + d|$ are *lengths*.

Looking ahead

As explained in Chapter III:

The theorems of complex numbers are basically the same as the corresponding theorems of real numbers. (Why?) Remember: The theorems of complex numbers are working "unlimited". Examples: We can operate with $\sqrt[n]{-a}$, where a is a real number > 0, and $n = 2, 4. 6, 8, \ldots$. *The Fundamental Theorem of Algebra* (FTA) states that every polynomial equation of degree n with complex coefficients has n roots in the complex-number system if each solution (root) is counted up to its *multiplicity.*

Therefore, of course, we do not need to directly prove, say, the ***binominal theorem for complex numbers***:

$$(z+w)^n = z^n + \binom{n}{1}z^{n-1}w + \binom{n}{2}z^{n-2}w^2 + \ldots + \binom{n}{n}w^n$$

where z and w are complex numbers, n is a positive integer

and $\binom{n}{r} = \dfrac{n!}{r!(n-r)!}$.

Taylor's formula and special Taylor series such as, say, $\dfrac{1}{1+z} = 1 + z + z^2 + z^3 + \ldots$ ($|z| <$

1), $e^z = 1 + z + \dfrac{z^2}{2!} + \dfrac{z^3}{3!} + \ldots$, are *as in real calculus*, with the real x replaced by complex z.

And, of course, we do have ***functions of a complex variable (and of several complex variables)***. A ***complex function*** f (z), where $z = x + iy$, is equivalent to a pair of *real* functions $u(x,y)$ and $v(x,y)$, each depending on the real variables x and y. If $z = x + iy$ and $w = u + iv$, then $w = $ f $(z) = u(x,y) + iv(x,y)$. Here z is called a ***complex variable***. Differentiable functions f (z) are called ***analytic functions***. Any f (z) which takes on a *single* definite value when a value is assigned to z, is called a *uniform function* of z. There is in general not a unique way (path) in which $\Delta z \to 0$. (See below.) But if f (z) is such that $f'(z_0)$ *is* the same for *all* paths by which $\Delta z \to 0$, then f (z) is said to be *monogenic* at z_0.

Uniformity and *monogenicity* are distinguishing features of *analytic* functions of a complex variable.

Example: $f(z) = z^2 + 2z$ is a complex function *defined* for all z; that is, its *domain* is the whole complex plane.

By the way, $f(z + 2n\pi i) = e^{z+2n\pi i} = e^z = f(z)$. Thus $f(z) = e^z$ has the period $2\pi i$.

Optional exercise: Show that $f(z + 2n\pi i) = f(z)$.

In the complex numbers we think of functions as maps of sets to sets, not numbers to numbers. Examples: $f(z) = z^2 = (-z)^2 = f(-z)$ is a two to-one function in the complex plane. $f(z) = z^k$ is a k-to-one function in the complex plane and $g(z) = z^{\frac{1}{k}}$ is a one-to-k function in the complex plane, where $k = 1, 2, 3, \ldots$.

A real-valued function of a real variable is often represented by the *graph* of the function, but when $w = f(z)$, where z and w are complex numbers, no such graph of $f(z)$ is available because z and w are located in a plane, not on a line. Instead the points $z = f(x, y)$ are indicated in a z plane, and the points $w = f(u, v)$ are indicated in a w plane. This is often referred to as a *mapping*. The *image* in the w plane of a point z (in the z plane) is $w = f(z)$.

The mapping of the z plane onto the w plane by $w = z + C$, where C is a *complex constant*, is a *translation* by means of the *vector* representing C. The mapping $w = Bz$, where B is a nonzero *complex constant*, is an *expansion* or *contraction* of the *radius vector* for z by the factor $|B|$ together with a *rotation* through an angle arg B about the origin. A general *linear transformation* $w = Bz + C$, which is a composition of the $Z = Bz$ $(B \neq 0)$ and $w = Z + C$, is an *expansion* or *contraction*, and a *rotation*, followed by a *translation*.

The laws of **limits and continuity** of functions of a complex variable are similar to the corresponding laws of limits and continuity of functions of a real variable. Why? Well, limits and continuity are basically multiplication, addition, division, and subtraction, plus limiting processes. By the way: The n zeroes of a polynomial of degree $n \geq 1$ with *complex coefficients* depend *continuously* upon the coefficients.

Examples:

$$\lim f(z) = w_0$$
$$z \to z_0.$$

$$\lim \frac{1}{z^2} = 0$$
$$z \to \infty.$$

Also, the laws of ***differentiation of functions of a complex variable*** are similar to the corresponding laws of differentiation of functions of a real variable. Why? Well, derivatives are basically multiplication, addition, division, and subtraction, plus limiting processes.

Examples:

$$f'(z_0) = \lim \frac{f(z_0 + \Box z) - f(z_0)}{\Box z}$$
$$\Box z \to 0$$

where $\Box z = z - z_0$.

So, if $f(z) = z^2$, then

$$f'(z_0) = \lim \frac{f(z_0 + \Box z) - f(z_0)}{\Delta z} = \lim \frac{(z_0 + \Box z)^2 - z_0^2}{\Delta z}$$
$$\Box z \to 0 \qquad\qquad \Box z \to 0$$
$$= \lim (2z_0 + \Delta z) = 2z_0$$
$$\Box z \to 0.$$

$$\frac{d}{dz}c = 0, \quad \frac{d}{dz}z = 1, \quad \frac{d}{dz}\big[cf(z)\big] = cf'(z),$$

$$\frac{d}{dz}z^n = nz^{n-1}, \quad \frac{d}{dz}\big[f(z) \pm F(z)\big] = f'(z) \pm F'(z),$$

$$\frac{d}{dz}\big[f(z)F(z)\big] = f(z)F'(z) + f'(z)F(z),$$

$$\frac{d}{dz}\left[\frac{f'(z)}{F'(z)}\right] = \frac{f'(z)F(z) - f(z)F'(z)}{\big[F(z)\big]^2},$$

$$\frac{d}{dz}e^z = e^z, \quad \frac{d}{dz}\sin z = \cos z, \quad \frac{d}{dz}\cos z = -\sin z.$$

Of course, the function $z(t) = x(t) + iy(t)$, where t is a real *parameter*, represents a **curve**

in the complex plane. If $z(t) = x(t) + iy(t)$, then $\dfrac{dz(t)}{dt} = \dfrac{dx(t)}{dt} + i\dfrac{dy(t)}{dt}$.

Integrals are basically mostly addition, so, again, the rules are the same:

Examples:

$$\int_{z_1}^{z_2} f(z)dz = F(z)\Big|_{z_1}^{z_2} = F(z_2) - F(z_1)$$

where $F'(z) = f(z)$.

Note: Integrals and antiderivatives are path-independent.

If $w(t) = u(t) + [v(t)]\mathbf{i}$, then $\displaystyle\int_a^b w(t)dt = \int_a^b u(t)\,dt + i\int_a^b v(t)\,dt$.

If $f(x) = e^{tx}$, where x is any real number, and t is a fixed complex number, then
$f'(x) = te^{tx}$,

as expected. Also, $\int e^{tx} dx = \dfrac{e^{tx}}{t}$, as expected.

$$\int_{i}^{i+1} z^2 dz = \frac{z^3}{3}\bigg|_{i}^{1+i} = \frac{(1+i)^3 - i^3}{3} = -\frac{2}{3} + i.$$

Differential equations; an example:

If $e^{\theta i}$ is a complex number, then $e^{\theta i} = A(\theta) + \big[B(\theta)\big]i$ where $A(\theta)$ and $B(\theta)$ are real-valued functions. Assuming $A(\theta)$ and $B(\theta)$ are differentiable, we have $\left(e^{\theta i}\right)i = A'(\theta) + \big[B'(\theta)\big]i$ and $-e^{\theta i} = A''(\theta) + \big[B''(\theta)\big]i$. Comparison gives $A''(\theta) = -A(\theta)$ and $B''(\theta) = -B(\theta)$. So $A''(\theta) + A(\theta) = 0$ and $B''(\theta) + B(\theta) = 0$. Each of the functions $A(\theta)$ and $B(\theta)$ is a solution of the differential equation $f'' + f = 0$. From calculus we know that this equation has exactly one solution with specified initial values $f(0)$ and $f'(0)$. If $\theta = 0$, we have $e^{0i} = 1$ and $-e^{0i} = -1$, so we must have $A(0) = 1,\ A'(0) = 0,\ B(0) = 0,\ B'(0) = 1$.

By the **uniqueness theorem for second-order differential equations with constant coefficients** we must have

$A(0) = \cos\theta,$

$B(0) = \sin\theta,$

and $e^{\theta i} = \cos\theta + (\sin\theta)i.$

Remember that $e^{xi} = \cos x + (\sin x)i$, where x is a real number, and that the generalization of this theorem is $e^{zi} = \cos z + (\sin z)i$, where z is a complex number.

Remember also: The generalization of de Moivre's theorem: If z and w are complex numbers, then $\big[\cos z + (\sin z)i\big]^w = \cos(wz) + \big[\sin(wz)\big]i$. But now $\big[\cos z + (\sin z)i\big]^w$ is a multivalued function while $\cos(wz) + \big[\sin(wz)\big]i$ is not. Therefore $\cos(wz) + \big[\sin(wz)\big]i$ is one value of $\big[\cos z + (\sin z)i\big]^w$.

Trigonometric functions and series; the story is the same. The view is clear. The complex-
number system is a perfect system. Examples:

From $e^{\theta i} = \cos\theta + (\sin\theta)i$ and $e^{-\theta i} = \cos\theta - (\sin\theta)i$ follows

$$\cos\theta = \frac{e^{\theta i} + e^{-\theta i}}{2},$$

$$\sin\theta = \frac{e^{\theta i} - e^{-\theta i}}{2i}.$$

Remember from APPENDIX 1:

$$e^{\theta} = \frac{\theta^0}{0!} + \frac{\theta^1}{1!} + \frac{\theta^2}{2!} + \frac{\theta^3}{3!} + ... + \frac{\theta^n}{n!} + ...,$$

$$e^{z} = \frac{z^0}{0!} + \frac{z^1}{1!} + \frac{z^2}{2!} + \frac{z^3}{3!} + ... + \frac{z^n}{n!} + ...,$$

$$\cos\theta = \frac{\theta^0}{0!} - \frac{\theta^2}{2!} + \frac{\theta^4}{4!} - ...,$$

$$\sin\theta = \frac{\theta^1}{1!} - \frac{\theta^3}{3!} + \frac{\theta^5}{5!} - ....$$

It can be shown that

$$\frac{z^0}{0!} - \frac{z^2}{2!} + \frac{z^4}{4!} - ... = \frac{e^{zi} + e^{-zi}}{2},$$

$$\frac{z^1}{1!} - \frac{z^3}{3!} + \frac{z^5}{5!} - ... = \frac{e^{zi} - e^{-zi}}{2i}.$$

And it can be shown that if

$$\cos z = \frac{z^0}{0!} - \frac{z^2}{2!} + \frac{z^4}{4!} - ... = \frac{e^{zi} + e^{-zi}}{2},$$

$$\sin z = \frac{z^1}{1!} - \frac{z^3}{3!} + \frac{z^5}{5!} - ... = \frac{e^{zi} - e^{-zi}}{2i},$$

then

$$\cos(-z) = \cos z,$$

$$\sin(-z) = -\sin z,$$

$$\sin\left(z + \frac{\pi}{2}\right) = \cos z,$$

$$\sin 2z = 2\sin z \cos z,$$

$$\cos 2z = \cos^2 z - \sin^2 z,$$

$$\cos^2 z + \sin^2 z = 1 \quad (\text{proof: } 1 = e^{iz}e^{-iz} = (\cos z + i\sin z)(\cos z - i\sin z) = \cos^2 z + \sin^2 z = 1),$$

$$\cos(z + w) = \cos z \cos w - \sin z \sin w,$$

$$\sin(z + w) = \sin z \cos w + \cos z \sin w,$$

$$\tan z = \frac{\sin z}{\cos z},$$

$$\cot z = \frac{\cos z}{\sin z},$$

$$\frac{d}{dz}\sin z = \cos z,$$

$$\frac{d}{dz}\cos z = -\sin z,$$

etc.

Obviously, all in all, it is *discovered* that

$$\cos z = \frac{e^{zi} + e^{-zi}}{2},$$

$$\sin z = \frac{e^{zi} - e^{-zi}}{2i}.$$

are fitting perfectly into the non-material world of mathematics.

Let $\sin z = \dfrac{e^{iz} - e^{-iz}}{2i} = x$, then $\arcsin\left(\dfrac{e^{iz} - e^{-iz}}{2i}\right) = z$. And let $k = e^{iz}$. Then we have

$\dfrac{k - \dfrac{1}{k}}{2i} = x$, and $k^2 - 2ikx - 1 = 0$. So $k = ix \pm \sqrt{1 - x^2} = e^{iz}$, $iz = \ln\left(ix + \sqrt{1 - x^2}\right)$

(+ is chosen), $z = \dfrac{1}{i}\ln\left(ix + \sqrt{1 - x^2}\right)$, and $z = -i\ln\left(ix + \sqrt{1 - x^2}\right)$, where $x = \dfrac{e^{iz} - e^{-iz}}{2i}$.

Optional exercise:

Show that $\dfrac{e^{(a+bi)i} - e^{-(a+b)i}}{2i} = \sin a\left(\dfrac{e^b + e^{-b}}{2}\right) + \left[\cos a\left(\dfrac{e^b - e^{-b}}{2}\right)\right]i$.

Complex matrices are important in mathematics. **Complex matrix equations**, three examples:

$$\begin{bmatrix} 1 & i \\ -i & 1 \end{bmatrix} = \frac{1}{\sqrt{2}}\begin{bmatrix} 1 & i \\ i & 1 \end{bmatrix}\begin{bmatrix} 0 & 2i \\ 0 & 0 \end{bmatrix}\begin{bmatrix} 1 & i \\ i & 1 \end{bmatrix}^T, \quad \begin{bmatrix} 1 & 0 \\ 0 & -1 \end{bmatrix} = \begin{bmatrix} 1 & 0 \\ 0 & 1 \end{bmatrix}\begin{bmatrix} 1 & 0 \\ 0 & i \end{bmatrix}^2, \quad A = B + \mathbf{i}C.$$

Defining a complex matrix in terms of a *power series*, example:

$A = I + iB + \dfrac{(iB)^2}{2!} + \dfrac{(iB)^3}{3!} + \dots$, where A is the $n \times n$ matrix $A = e^{iB}$, e is *Euler's number*, $\mathbf{i}$ is the *imaginary unit*, I is the *identity matrix*, and B is an $n \times n$ matrix.

For the **matrix function** $A(t) = \left(a_{ij}(t)\right)$, where (here) A is a complex matrix, we define the derivative $\dfrac{dA}{dt}$ to be the matrix function $\dfrac{dA}{dt} = \left(\dfrac{da_{ij}}{dt}\right)$. We define $\int A(t)dt$ to be the matrix $\int A(t)dt = \left(\int a_{ij}(t)dt\right)$ and $\int_a^b A(t)dt$ to be the matrix $\int_a^b A(t)dt = \left(\int_a^b a_{ij}(t)dt\right)$.

Twistor space: The *twistor space* is defined by four complex dimensions. Since a *complex number* consists of two independent parts ($z = a + bi = (a, b)$), it contains more information than the conventional space-time with four real dimensions (x, y, z, t). *Twistor theory* can vastly simplify calculations needed in advanced physics: "Things that were much too hard to do even with a supercomputer, now you need just a good postdoc and a piece of paper." Zvi Bern.

Software:

A complex number computer to multiply and divide complex numbers was developed and built during 1938 – 1940 by Bell Telephone laboratories. It was used 1940 – 1949 on telephone switching problems, etc.

Remember: There are free *complex number calculators* on the Internet. And there exist *pocket calculators* supporting complex numbers. EXCEL has several functions to support complex numbers math. The most advanced and popular software packages are: *Mathematica* (fully integrated symbolic and numerical software for scientific and technical computing), *Maple* (math and engineering software), *Matlab* (technical computing software, uses Maple as its 'Symbolic Toolkit'), *Mathcad* (engineering calculation software).

We also have: Lascaux Software's: $f(z)$ – *The Complex Variables Program*. "f (z) is Windows program that puts the world of complex variables at your fingertips – with absolutely no programming."

Applications of complex numbers

A mathematical papyrus from the ancient Egypt offers the earliest known occurrence of the square root of a negative number. This papyrus uses the square root of minus one to calculate the volume of a truncated square pyramid. As we know, after hundreds of years of bitter debates the complex numbers finally won acceptance and were put into use in mathematics and physics in Napoleonic times. Already in 1823 Fresnel used complex numbers in his theory of total reflexion. The Universe is described by Relativity Theory and Quantum Theory. Relativity Theory uses complex numbers, and Quantum Theory uses complex numbers for just about everything. Actually, as we know, there is no way one can formulate quantum mechanics without resorting to complex numbers. The calculus of functions of a complex variable has become an indispensable tool for physicists and engineers.

While any object contains measurable parameters, it also has numerous essences that are not observable or computable. Imaginary numbers may represent these essences, qualitative aspects, or potentialities that in turn can evolve the object to a new and measurable state. We may take the real numbers to represent **observables** and interpret imaginary numbers as **non-observables**. Complex numbers don't appear in measurements. They are merely a useful bookkeeping tool for tracking the phase of a quantity. Nothing measurable ever contains **i** (the imaginary unit).

The process described above is happening even in mathematics itself, as we already know. Example: The cubic equation $x^3 = 15x + 4$ has the real root (solution) $x = 4$. Tartaglia's cubic formula gives $x = \sqrt[3]{2 + \sqrt{-121}} + \sqrt[3]{2 - \sqrt{-121}}$, but $\left(2 \pm \sqrt{-1}\right)^3 = 2 \pm \sqrt{-121}$, so $x = \left(2 + \sqrt{-1}\right) + \left(2 - \sqrt{-1}\right) = 4$.

The theory of *complex calculus* has applications to mathematical problems that at first do not seem to involve complex numbers. Example: The proof that $\int_0^\infty \frac{\sin^2 x}{x^2} dx = \frac{\pi}{2}$, or similar proofs, may be difficult or impossible using real calculus only, but can be readily proved using complex calculus.

No wonder a list of uses of complex numbers is long. Here is an *incomplete* list of uses (in alphabetical order):

Advanced calculus
advanced engineering mathematics
algebra (even basic algebra)
algebraic geometry
aeronautics
algebraic geometry
algebraic logic
analysis of stress on beams
analytic functions
analytic geometry
analyzing stresses and strains
animation of motion
applied and computational complex analysis
applied mathematics
boundary value problems
calculus
calculus of residues
Cauchy-Riemann equations
chaotic dynamical systems
characteristic functions
chemistry
chemometrics
circuits analysis
Clifford algebras
color coding a three-dimensional graph to suggest four dimensions
combine rotations (using multiplicatuion)
combine translations (using addition)
complex Abelian algebras
complex Abelian groups
complex algebra
complex algebraic geometry
complex algebraic spaces/manifolds
complex algebraic surfaces
complex algebraic topology
complex analysis
complex analysis applied to potential theory
complex analytic geometry
complex analytic spaces/manifolds

complex angles
complex arithmetic
complex Banach spaces/manifolds
complex calculus
complex derivative
complex differential equations
complex differential geometry
complex distances
complex dynamical systems
complex dynamics
complex eigenvalues
complex Euclidean space
complex Finsler spaces/manifolds
complex Floquet modes
complex Fourier analysis
complex Fourier series
complex Fourier transform
complex function theory
complex gauge theory
complex geometry
complex group theory
complex Hamiltonian spaces/manifolds
complex heat potential
complex hyperbolic spaces/manifolds
complex integral equations
complex integrals
complex interpolation of spectral segments
complex Lagrange spaces/manifolds
complex Lie algebra
complex Lie groups
complex linear algebra
complex linear spaces/manifolds
complex manifolds
complex matrices
complex matrix algebra
complex matrix analysis
complex matrix arithmetic
complex matrix calculus
complex matrix equations
complex multilinear algebra
complex Navier-Stokes equations

complex Noetherian spaces
complex non-hyperbolic spaces/manifolds
complex non-linear equations
complex non-linear integral equations
complex number theory
complex operators
complex physical spaces
complex polynomial equations
complex polynomials
complex polytopes
complex potential
complex probability theory
complex quantum mechanics (CQM)
complex rays
complex Snell law
complex spaces
complex space-time
complex structures
complex Taylor series
complex topology
complex trigonometric functions
complex vector spaces
complex vectors
computer algebra system
computer graphics
construction of regular polygons
contour integration (methods of contour integration)
control theory
DCT to decode images
describing radiation
describing rotations (by using multiplication)
describing wave motion
design of dynamos
design of electric motors
designing tuned radio circuits
differential equations
digital image processing
Dirac matrices in quantum mechanics
Dirichlet problem
DSP (Digital Signal Processing)
elasticity

electric fields

electrical engineering: voltage, current, resistance, inductance, impedance, capacitance, frequency, phase differences

electromagnetics

electromagnetism

electronic circuit design

electronics

elliptic functions with applications

engineering electromagnetics

engineering field problems

experimental data estimation

fast Fourier transform

field mapping

Finsler geometries

flow around a corner

flow around a cylinder

flow of fluids

fluid dynamics

fluid mechanics

Fourier analysis

Fourier series

Fourier theory

Fourier transformations

fractal geometry

fractal images

fractals

frequency-domain analysis of functions

Fresnel's theory of total reflexion

Gaussian integers

general relativity

generalization of polyhedra

geometric algebra

geometric analysis

geometric calculus

geometrical applications

geometry

Hamilton geometries

harmonic functions

heat conduction

Heisenberg's matrix mechanics

Heron's formula via complex numbers

higher-dimensional complex spaces
Hilbert spaces
hydrodynamics
hyperbolic complex numbers
hypercomplex numbers
imaginary number
imaginary part
improper integrals
infinite-dimensional complex spaces
infinite series
interference experiments (in quantum theory)
investigation of electrical currents
investigation of wavelength
Julia sets
Lagrange geometries
Laplace transform
laws of planetary motions
light-cone geometry (in relativity)
logarithms
magnetic fields
making some integrals simpler
Mandelbrot set
manipulating *two* variables as a *single* quantity
manipulation of large matrices used in modelling
matrix analysis
matrix multiplication
matrix operations for image processing
mechanical analog of electrical circuits
modelling aeroplane wings
modelling the weather
movement of chock absorbers in cars
multivariate impedance measurements
neural networks
Nichols plot
nuclear magnetic resonance
number theory
Nyquist plot
octonion(ic) quantum mechanics (OQM)
octonion(ic) algebra
octonion(ic) analysis
octonion(ic) arithmetic

octonion(ic) functions
octonion(ic) geometry
octonion(ic) manifolds
octonion(ic) space
operational mathematics
optimization theory
oscillations and waves
physics
plane geometry
Poisson's Integral Formula
potential theory
processing digital audio signals
processing digital video signals
proofs in number theory
proofs in real analysis
quantum chemistry
quantum field theory
quantum mechanics
quantum theory
quaternion(ic) algebra
quaternion(ic) analysis
quaternion(ic) arithmetic
quaternion(ic) calculus
quaternion(ic) functions
quaternion(ic) geometry
quaternion(ic) quantum mechanics (QQM)
quaternion(ic) theory (complex numbers are a subset of quaternions)
quaternion(ic) spaces
quaternion(ic) variables
represent the complete state of a solid object
representation of digital information
representing rotations
resonance phenomena
Riemann surfaces
Schroedinger's equation
series representations
signal analysis
simplify calculations involving trigonometric functions
singular complex spaces/manifolds
solving differential equations (even finding real solutions)
solving polynomial equations

solving some integrals
solving the cubic equation (even when all 3 roots are real)
special relativity
spherical trigono-geometry
spinor calculus
spinor field theory
spinors (generalization of tensors)
statistical models
Stein space
surveying (representing line segments)
Tartaglia's cubic formula
temperature distribution
theoretical hydrodynamics
theory of elasticity
topology
transcomplex numbers
transformational geometry
transmission of digital information
twistor algebra
twistor analysis
twistor arithmetic
twistor calculus
twistor functions
twistor geometry
twistor manifolds
twistor spaces
twistor theory
two dimensional rotations
vector analysis
vector analysis in electronics
vectors in two dimensions
vibration analysis
vortices of flow
wave motion analysis
z-transform

BIBLIOGRAPHY (BOOKS)

Abramowitz, M., and A. Stegun (eds.): **Handbook of Mathematical Functions**

Adam, J. A.: **A Mathematical Nature Walk**

Adams, C., et al: **How to Ace Calculus**

Adams, C., et al: **How to Ace the Rest of Calculus**

Adams. R. A.: **Sobolev Spaces**

Adler, I: **The New Mathematics**

Ahlfors, L. V.: **Complex Analysis**

Akopyan, A. V., and A. A. Zaslavsky: **Geometry of Conics**

Aleksandrov, A. D.: **Mathematics: Its Content, Methods and Meaning**

Alexander H, and J. Wermer: **Several Complex Variables and Banach Algebras**

Alexander, R. D.: **Real Transformations From Complex Numbers**

Allenby, R.: **Numbers and Proofs**

Anderson, J. W.: **Hyperbolic Geometry**

Andes, J., and L. Górniewicz: **Infinite Dimensional Analysis: Hitchhiker's Guide**

Andreescu, T., and D. Andrica: **Complex Numbers from A to … Z**

Andreescu, T., et al.: **104 Number Theory Problems: From the Training of the USA IMO Team**

Andreescu, T.: **Mathematical Olympiad Challenges**

Anglin, W. S.: **Mathematics. A Concise History and Philosophy**

Anton, H., and C. Rorres: **Elementary Linear Algebra. Applications Version**

Apostol. T. M.: **Calculus** (Two volumes)

Apostol, T. M.: **Early History of Mathematics**

Argand, J. R.: **Essay on a Method of Representing Imaginary Quantities** (Published1813)

Atkinson, F. V.: **On Sums and Powers of Complex Numbers: An Improved Estimate**

Aubin, J.-P., and H. Frankowska: **Set-Vaued Analysis**

Avigad, J.: **Number Theory and Elementary Arithmetic**

Awodey, S.: **Category Theory**

Axler, S.: **Linear Algebra Done Right**

Backhouse, J. K., et al.: **Pure Mathematics** (Two volumes)

Bailey, D. H., et al.: **Experimental Mathematics in Action**

Bak, J., and D. J. Newman: **Complex Analysis**

Balaguer, M.: **Platonism and Anti-Platonism in Mathematics**

Barnett, S.: **Introduction to Mathematical Control Theory**

Barrow, J. D.: **One Hundred Essential Things You Didn't Know You Didn't Know: Math Explains Your World**

Baumslag, B.: **Schaum's Outline of Group Theory**

Beckmann, P.: **A History of Pi**

Bell, E. T.: **Men of Mathematics**

Bellos, D., et al.: **The Universal History of Numbers**

Berlekamp, E. R., et al.: **Winning Ways For Your Mathematical Plays**

Bettye, A. C., and A. M. Leggett: **Complexities: Women in Mathematics**

Bhardwaj, D.: **Complex Numbers Made Easy**

Birkenhake, C., and H. Lange: **Complex Abelian Varieties**

Bishop, R. L., and S. I. Goldberg: **Tensor Analysis on Manifolds**

Bloch, E. D.: **A First Course in Geometric Topology and Differential Geometry**

Bloch, E. D.: **Proofs and Fundamentals: A First Course in Abstract Mathematics**

Bold, B.: **Famous Problems of Geometry and How to Solve Them**

Bourbaki, N.: **General Topology: Chapters 5 – 6**

Bourbaki, N.: **Lie Groups and Lie Algebras: Chapters 1-3**

Boyer, C. B., et al.: **A History of Mathematics**

Braun, M.: **Differential Equations and Their Applications**

Bray, D.: **Wetware: A Computer in Every Living Cell**

Bredon, G. E.: **Sheaf Theory**

Brenner, S. C.: **The Mathematical Theory of Finite Element Methods**

Bressoud, D. M.: **Second Year Calculus**

Britton, N. F.: **Essential Mathematical Biology**

Bronson, R.: **Operations Research**

Brookshear, J. G.: **Computer Science: An Overview**

Bryant, V.: **Yet Another Introduction to Analysis**

Bueng, O., and O. Linnebo: **New Waves in Philosophy of Mathematics**

Buss, S. R.: **Handbook of Proof Theory**

Byers, W.: **How Mathematicians Think**

Cajori, F.: **A History of Mathematical Notations** (Two volumes)

Callahan, J. J.: **The Geometry of Spacetime**

Carico, C. C.: **Complex Numbers; polynomial functions**

Carr, V.: **Complex Numbers Made Simple**

Carter, P. J.: **The Complete Book of Intelligence Tests**

Catoni, F, et al.: **The Mathematics of Minkowski Space-Time: With an Introduction to Commutative Hypercomplex Numbers**

Childs, L.: **A Concrete Introduction to Higher Algebra**

Clawson, C. C.: **Mathematical Mysteries: The Beauty and Magic of Numbers**

Clegg, B.: **A Brief History of Infinity**

Cockcraft, W. H.: **Complex Numbers: A Study in Algebraic Structures**

Coexter, H. S. M.: **Non-Euclidean Geometry**

Coexter, H. S. M.: **Projective Geometry**

Coexter, H. S. M.: **Regular Complex Polytopes**

Coexter, H. S. M.: **Regular Polytopes**

Coexter, H. S. M.: **The Beauty of Geometry: Twelve Essays**

Cohen, D. I. A.: **Introduction to Computer Theory**

Cohn, M.: **An Introduction to Ring Theory**

Conway, J. B.: **Functions of One Complex Variable**

Conway, J. H., and D. Smith: **On Quaternions and Octonions**

Conway, J. H.: **On Numbers and Games**

Conway, J. H.: **The Symmetries of Things**

Corrochano, E. B., and G. Sobczyk: **Geometric Algebra**

Courant, R., et al.: **What Is Mathematics?**

Cover, T. M., and J. A. Thomas: **Elements of Information Theory**

Cox, D., et al.: **Ideals, Varieties, and Algorithms**

Darling, D.: **The Universal Book of Mathematics**

Das, A.: **Tensors: The Mathematics of Relativity Theory and Continuum Mechanics**

Davis, M. D.: **Game Theory: A Nontechnical Introduction**

Deaux, R.: **Introduction to the Geometry of Complex Numbers**

Dedekind, R.: **What Are Numbers, and What Do They Mean?**

Dehaene, S.: **The Number Sense**

Derbyshire, J., and J. H. Press: **Unknown Quantity: A Real and Imaginary History of Algebra**

Devaney, R. L.: **An Introduction to Chaotic Dynamical Systems**

Devaney, R. L., and L. Keen: **Complex Dynamics**

Devlin, K.: **Mathematics: The New Golden Age**

Devlin, K.: **Sets, Functions, and Logic: An Introduction to Abstract Mathematics**

Devlin, K.: **The Language of Mathematics: Making the Invisible Visible**

Devlin, K.: **The Math Gene**

Dienes, P.: **The Taylor Series**

Dixon, G. M.: **Division Algebras: Octonions, Quaternions, Complex Numbers, and the Algebraic Design of Physics**

Dodson, C. T. J., and T. Poston: **Tensor Geometry**

Dorrie, H.: **100 Great Problems of Elementary Mathematics**

Dunham, W.: **Journey Through Genius: The Great Theorems of Mathematics**

Eddington, A. S.: **The Mathematical Theory of Relativity**

Edelstein-Keshet, L.: **Mathematical Models in Biology**

Eisenhart, L. P.: **An Introduction to Differential Geometry – With the Use of Tensor Calculus**

Engel, A.: **Problem-Solving Strategies**

Erdmann K., and M. J. Wildon.: **Introduction to Lie Algebras**

Ernest, P.: **The Philosophy of Mathematics Education**

Esposito, G.: **Complex General Relativity**

Estermann, T.: **Complex Numbers and Functions**

Euclid and Andrew Aberdein: **The Elements: Books I-XIII – Complete and Unabridged**

Evans, L.: **Complex Numbers and Vectors**

Eves, H., **Foundations and Fundamental Concepts of Mathematics**

Faber, R. L.: **Differential Geometry and Relativity Theory: An Introduction**

Faires, J. D.: **First Steps for Math Olympians: Using the American Mathematics Competitions**

Farmelo, G.: **It Must Be Beautiful: Great Equations of Modern Science**

Fauvel, J., et al.: **Music and Mathematics: From Pythagoras to Fractals**

Fischer, G.: **Complex Analytic Geometry**

Fleisch, D.: **A Student's Guide to Maxwell's Equations**

Frascolla, P.: **Wittgenstein's Philosophy of Mathematics**

Frege, G.: **The Foundations of Arithmetic**

Gardner, M.: **Entertaining Mathematical Puzzles**

Gardner, M.: **Martin Gardner's Table Magic**

Gardner, M.: **My Best Mathematical and Logic Puzzles**

Gardner, M.: **Perplexing Puzzles and Tantalizing Teasers**

Gardner, M.: **The Colossal Book of Mathematics: Classic Puzzles, Paradoxes, and Problems**

Gardner, M.: **The Colossal Book of Short Puzzles and Problems**

Garfunkel, J.: **Solving Problems in Geometry By Using Complex Numbers**

Gass, S. I.: **An Illustrated Guide to Linear Programming**

Gelfand, I. M.: **Calculus of Variations**

Gelfand, I. M., et al: **Functions and Graphs**

Gibilisco, S.: **Math Proofs Demystified**

Gillies, D.: **Philosophical Theories of Probability**

Gilmore, R.: **Lie Groups, Lie Algebras, and Some of Their Applications**

Gohlberg, I, and S. Goldberg: **Basic Operator Theory**

Gowers, T., et al. (eds): **The Princeton Companion to Mathematics**

Gray, J.: **Worlds Out of Nothing**

Gregg, J. R.: **Ones and Zeros: Understanding Boolean Algebra, Digital Circuits, and the Logic of Sets**

Greub, W. H.: **Multilinear Algebra**

Griffiths, D. J.: **Introduction to Electrodynamics**

Griffiths, D. J.: **Introduction to Quantum Mechanics**

Grimmett, G., and D. Stirzaker: **One Thousand Exercises in Probability**

Guillemin. V., and A. Pollack: **Differential Topology**

Gurney, K.: **An Introduction to Neural Networks**

Hahn, L. S.: **Complex Numbers and Geometry**

Halmos, P. R.: **A Hilbert Space Problem Book**

Halmos, P. R.: **Measure Theory**

Halmos, P. R.: **Naïve Set Theory**

Hamilton, A. G.: **Logic for Mathematicians**

Hamilton, A. G.: **Numbers, Sets and Axioms: The Apparatus of Mathematics**
Hamming, R. W.: **Numerical Methods for Scientists and Engineers**
Hanselman, D. C.: **Mastering Mathlab 7**
Hanson, A.: **Visualizing Quaternions**
Hardy, G. H.: **A Course of Pure Mathematics**
Hardy, G. H.: **A Mathematician's Apology**
Hardy, G. H.: **Course of Pure Mathematics**
Hardy, G. H., et al.: **An Introduction to the Theory of Numbers**
Harris, J., et al.: **Combinatorics and Graph Theory**
Harrison, C. W.: **Tables for the Square Root of Complex Numbers**
Harrison, J.: **Handbook of Practical Logic and Automated Reasoning**
Hart, W. D.: **The Philosophy of Mathematics**
Hartshorne, R.: **Geometry: Euclid and Beyond**
Havil, J., and F. Dyson: **Gamma: Exploring Euler's Constant**
Havil, J.: **Nonplussed: Mathematical Proof of Implausible Ideas**
Hawkins, F. M., and J. Q. Hawkins: **Complex Numbers and Elementary Complex Functions**
Hellman, H.: **Great Feuds in Mathematics**
Henrion, C.: **Women in Mathematics**
Herman, J., et al.: **Equations And Inequalities: Elementary Problems And Theorems In Algebra and Number Theory**
Hersh, R.: **What Is Mathematics, Really?**
Heyting, A.: **Intuitionism, An Introduction**
Hilbert, D.: **Geometry and Imagination**
Hilbert, D.: **Principles of Mathematical Logic**
Hilbert, D.: **The Foundations of Geometry**
Hildebrand, F. B.: **Finite-Difference Equations and Simulations**
Hille, E.: **Analytic Function Theory** (Two volumes)
Hime, H. W. L.: **The Outlines of Quaternions**
Hirose, A.: **Complex-Valued Neural Networks**
Holmes, M. H.: **Introduction to Perturbation Methods**
Hoppensteadt, F. C.: **Mathematics in Medicine and the Life Sciences**
Horn, R. A., and C. R. Johnson: **Matrix Analysis**
Houston, K.: **How to Think Like a Mathematician**
Howard, E.: **Foundations and Fundamental Concepts of Mathematics**
Howie, J. M.: **Complex Analysis**
Huggett, S. A., and K. P. Tod: **An Introduction to Twistor Theory**
Huizinga, J.: **Homo Ludens**
Huntley, H. E.: **The Divine Proportion: A Study of Mathematical Beauty**
It□, K.: **Encyclopedic Dictionary of Mathematics** (Four volumes)
Ivanov, O. A.: **Easy as Π? An Introduction to Higher Mathematics**

Jackson, J. D.: **Mathematics for Quantum Mechanics**

Jagerman, L. S.: **The Mathematics of Relativity for the Rest of Us**

Jain, M. C.: **Vector Spaces and Matrices in Physics**

James, J. F.: **A Student's Guide to Fourier Transforms: With Applications in Physics and Engineering**

Jech, T. J.: **The Axiom of Choice**

Jordan, T. F.: **Linear Operators for Quantum Mechanics**

Jordan, T. F.: **Quantum Mechanics in Simple Matrix Form**

Jourdain, P. E. B.: **The Nature of Mathematics**

Kajiwara, J., et al.: **Finite or Infinite Dimensional Complex Analysis**

Kalman, K.: **The Complex Number System (School Mathematics Study Group)**

Kantor, I. L., and A. S. Solodovnikov: **Hypercomplex Numbers: An Elementary Introduction to Algebras**

Karam, P. A.: **The Growing Use of Complex Numbers in Mathematics**

Kaplan, D.: **Understanding Nonlinear Dynamics**

Kaplan, W.: **Advanced Calculus**

Karatsuba, A. A.: **Complex Analysis in Number Theory**

Kay, D. C.: **Tensor Calculus**

Kazdan, J. L.: **Lectures on Complex Numbers and Infinite Series**

Kelley, J. L.: **General Topology**

Kellogg, O. D.: **Foundation of Potential Theory**

Kenwal, R. P.: **Generalized Functions: Theory and Technique**

Kenyon, D., and P. Grosshauser: **History of Complex Numbers (Grades 6 – 12)**

King, J. P.: **Mathematics in 10 Lessons: The Grand Tour**

Klein, F.: **Elementary Mathematics From An Advanced Standpoint**

Klein, M.: **Mathematical Thought from Ancient to Modern Times**

Kleinrock, L., and R. Gail: **Queueing Systems: Problems and Solutions**

Kline, M.: **Mathematics and the Physical World**

Knight, B., and R. Adams: **Complex Numbers and Differential Equations**

Knuth, D. E.: **Surreal Numbers**

Knuth, D. E.: **The Art of Computer Programming** (Three volumes)

Kodaira, K.: **Complex Manifolds and Deformation of Complex Structures**

Kogbetliantz, E. G.: **Handbook of First Complex Prime Numbers**

Krantz, S. G.: **Function Theory of Several Complex Variables**

Krantz, S. G.: **Partial Differential Equations and Complex Analysis**

Kreyszig, E.: **Introductory Functional Analysis with Application**

Kumbar, M. M.: **Mathematical Methods For Chemistry Beginners**

Kunen, K.: **The Foundations of Mathematics (Logic)**

Lacatos, I., et al.: **Proofs and Refutations: The Logic of Mathematical Discovery**

Laidlaw, D., and J. Weickert: **Visualization and Processing of Tensor Fields: Advances and Perspectives**

Lakoff, G., and R. Nunez: **Where Mathematics Comes From**

Landau, E.: **Foundations of Analysis**

Landman, B. M., and A. Robertson: **Ramsey Theory on the Integers**

Landsburg, S. E.: **The Big Questions**

Lang, S.: **Basic Mathematics**

Lang, S.: **Complex Analysis**

Lang, S.: **Introduction to Complex Hyperbolic Spaces**

Larson, L. C.: **Problem Solving Through Problems**

Lawler, G. F.: **Introduction to Stochastic Processes**

Levi, M.: **The Mathematical Mechanic: Using Physical Reasoning to Solve Problems**

Levi-Civita, T.: **The Absolute Differential Calculus (Calculus of Tensors)**

Levy, A.: **Basic Set Theory**

Liebeck, M.: **A Concise Introduction to Pure Mathematics**

Linderholm, C. E.: **Mathematics Made Difficult**

Lipschutz, S.: **Essential Computer Mathematics**

Lipschutz, S., et al.: **Complex Variables**

Livio, M.: **Is God a Mathematician?**

Lockhart, P., and K. Devlin: **A Mathematician's Lament: How School Cheats Us Out of Our Most Fascinating and Imaginative Art Form**

Longo, F.: **Absolutely Nasty Sudoku Level 4 (Mensa)**

Lorenz, E. N.,: **The Essence of Chaos**

Loucks, S. E.: **Introductory Physics with Algebra**

Lounesto, P.: **Clifford Algebra and Spinors**

Lovelock, D., and H. Rund: **Tensors, Differential Forms, and Variational Principles**

Lovitt, W. V.: **Linear Integral Equations**

Lutzen, J. (ed.): **Around Caspar Wessel and the Geometric Representation of Complex Numbers**

Lyusternik, L. A.: **Ten-Decimal Tables of the Logarithms of Complex Numbers**

Macilwaine, P. S., et al.: **Coordinate Geometry and Complex Numbers**

Maier, E.: **Modeling Real and Complex Numbers**

Maor, E.: **"e": The Story of a Number**

Marar, W. L.: **Real and Complex Singularities**

Markushevich, A. I.: **Complex Numbers and Conformal Mapping**

Marsden, J. E.: **Basic Complex Analysis**

Martin, D.: **Solving Problems in Complex Numbers**

Martin, G. E.: **Transformation Geometry**

Martzloff, J.-C.: **A History of Chinese Mathematics**

Matthews, P. C.: **Vector Calculus**

Mauro, B.: **Twenty-seven Uses for Imaginary Numbers**

Mazur, B.: **Imagining Numbers**

McBrewster, J., et al.: **Euler's Formula**

McBrewster, J., et al.: **Mathematics of General Relativity**

McBrewster, J.: **Natural Logarithm**

McMahon, D.: **Complex Variables Demystified**

Megginson, R. E.: **An Introduction to Banach Space Theory**

Mendelson, E.: **Introduction to Mathematical Logic**

Milewski, E. G. (Chief Editor): **Complex Variables Problem Solver**

Miller, F. P., et al.: **Exterior Algebra**

Miller, F. P., et al.: **Imaginary Unit**

Miller, F. P., and A. F. Vandome: **Mathematics of General Relativity**

Miller, F. P., et al.: **Metric Tensor**

Miller, K. S.: **Advanced Complex Calculus**

Milnor, J. W.: **Dynamics in One Complex Variable**

Mitchell, M.: **Complexity: A Guided Tour**

Mitchell, T. M.: **Machine Learning**

Mlodinow, L.: **The Drunkard's Walk: How Randomness Rules Our Lives**

Moivre, Abraham de: **Miscellanea Analytica** (London: 1730)

Moon, P. H., and D. E. Spencer: **Theory of Holors: A Generalization of Tensors**

Morris, D.: **Complex Numbers – the Higher Dimensional Forms**

Mosteller, F.: **Fifty Challenging Problems in Probability with Solutions**

Mujica, J.: **Compolex Analysis in Banach Spaces**

Naess, A., and O. Gron: **Introduction to General Relativity and its Mathematics** (Rapport 1998 Nr. 14)

Nahin, P. J.: **An Imaginary Tale: The Story of i**

Nahin, P. J.: **Chases and Escapes: The Mathematics of Pursuit and Evasion**

Nahin, P. J.: **Dr. Euler's Fabulous Formula: Cures Many Mathematical Ills**

Nahin, P. J.: **When Least Is Best: How Mathematicians Discovered Many Clever Ways to Make Things as Small (or as Large) as Possible**

Needham, T.: **Visual Complex Analysis**

Nelson, R. D. (ed): **The Penguin Dictionary of Mathematics**

Neumann, J. von: **Mathematical Foundations of Quantum Mechanics**

Newman, J. R.: **The World of Mathematics** (Four volumes)

Nickel, J.: **Mathematics: Is God Silent?**

Nuffield Foundations: **Nuffield Advanced Mathematics: Complex Numbers and Numerical Methods**

O'Donnell, P. J.: **Introduction to 2-Spinors in General Relativity**

Oeijord, N. K.: **The Very Basics of Tensors**

Olariu, S., et al.: **Complex Numbers in n Dimensions**

Olsen, S.: **The Golden Section: Nature's Greatest Secret**

O'Neill, B.: **Semi-Riemannian Geometry With Applications**

Palais, R. S.: **Foundations of Global Non-Linear Analysis**

Papastavridis, J. G.: **Tensor Calculus and Analytical Dynamics**

Pappas, T.: **More Joy of Mathematics: Exploring Mathematics All Around You**

Peitgen, H. O., and P. H. Richter: **The Beauty of Fractals: Images of Complex Dynamical Systems**

Penrose, R.: **The Road to Reality**

Petersen, K. E.: **Ergodic Theory**

Pickover, C. A.: **The Math Book: From Pythagoras to the 57th Dimension, 250 Milestones in the History of Mathematics**

Polster, B.: **Q. E. D.: Beauty in Mathematical Proof**

Polya, G.: **How to Solve It**

Polya, G., and J. Kilpatrick: **The Stanford Mathematics Problem Book: With Hints and Solution**

Ponnusamy, S., et al.: **Complex Variables with Applications**

Posamentier, A. S., and C. T. Salkind: **Challenging Problems in Algebra**

Posamentier, A. S.: **Challenging Problems in Geometry**

Posamentier, A. S.: **Math Wonders to Inspire Teachers and Students**

Posamentier, A. S.: **Mathematical Amazements and Surprises: Fascinating Figures and Noteworthy Numbers**

Pressley, A.: **Elementary Differential Geometry**

Prugovecki, E.: **Quantum Mechanics in Hilbert Space**

Rainich, G. Y.: **Mathematics of Relativity**

Ramsden, P, et al.: **Experiments in Undergraduate Mathematics: A Mathematica Based Approach**

Rashevsky, N.: **Mathematical Biology of Social Behavior**

Rashevsky, N.: **Mathematical Theory of Human Relations**

Reade, J., et al.: **Calculus with Complex Numbers**

Rendall, A. D.: **Partial Differential Equations in General Relativity**

Resnikoff, H. L., and R. O. Wells Jr.: **Mathematics in Civilization**

Richeson, D. S.: **Euler's Gem: The Polyhedron Formula and the Birth of Topology**

Robson, E.: **Mathematics in Ancient Iraq: A Social History**

Rosenhouse, J.: **The Monty Hall Problem: The Remarkable Story of Math's Most Contentious Brain Teaser**

Ross, S.: **A First Course in Probability**

Ross, K. A.: **Elementary Analysis: The Theory of Calculus**

Rotman, J., et al.: **Introduction to Homological Algebra**

Roy, S. C.: **Complex Numbers: Lattice Simulation and Zeta Function Applications**

Rudman, P. S.: **How Mathematics Happened: The First 50 000 Years**

Russell, B.: **Introduction to Mathematical Philosophy**

Russell, B.: **The Principles of Mathematics**

Safier, F.: **Precalculus**

Salzer, H. E.: **Table of Powers of Complex Numbers**

Sarason, D.: **Complex Function Theory**

Scheidemann, V.: **Introduction to Complex Analysis in Several Variables**

Schelkunoff, S. A.: **Complex Numbers in Elementary Mathematics**

Schertz, R.: **Complex Multiplication**

Schey, H. M.: **Div, Grad, Curl, and All That**

School Mathematics Project: **Complex Numbers**

School Mathematics Project: **Complex Numbers Unit Guide**

Schwerdtfeger, H.: **Geometry of Complex Numbers**

Seibel, P.: **Coders at Work**

Seife, C.: **Zero: The Biography of a Dangerous Idea**

Serre, J. P.: **A Course in Arithmetic**

Shakarchi, R.: **Problems and Solutions for Complex Analysis**

Shanker, S.: **Wittgenstein and the Turning Point in the Philosophy of Mathematics**

Shapiro, S.: **The Oxford Handbook of Philosophy of Mathematics and Logic**

Shaw, W. T.: **Complex Analysis with Mathematica**

Shklarsky, D. O.: **The USSR Olympiad Problem Book: Selected Problems and Theorems of Elementary Mathematics**

Sibley, T. Q.: **The Foundations of Mathematics**

Simmons, G. F.: **Precalculus Mathematics in a Nutshell**

Simmons, V.: **Pre-Math: The Success Training Program of Increasingly Complex Numbers Skills**

Smith, K. E., et al.: **An Invitation to Algebraic Geometry**

Sommerfeld, A.: **Partial Differential Equations in Physics**

Spain, B.: **Tensor Calculus: A Concise Course**

Spiegel, M. R.: **Advanced Mathematics** (Schaum's Outline)

Spiegel, M. R.: **Laplace Transforms** (Schaum's Outline)

Spiegel, M. R.: **Vector Analysis and an Introduction to Tensor Analysis**

Spivak, M.: **A Comprehensive Introduction to Differential Geometry** (Five volumes)

Spivak, M.: **Calculus**

Standish, R.: **Theory of Nothing**

Starzak, M. E.: **Mathematical Methods in Chemistry and Physics**

Stein, J. D.: **How Math Explains the World**

Stewart, I.: **Cabinet of Mathematical Curiosities**

Stewart, I.: **Galois Theory**

Stewart, I.: **Nature's Numbers: The Unreal Reality of Mathematics**

Stewart, I.: **Why Beauty Is Truth: A History of Symmetry**

Stickels, T. H.: **The Big Book of Mind-Bending Puzzles** (Mensa)

Stillwell, J.: **Numbers and Geometry**

Strogatz, S. H.: **Nonlinear Dynamics And Chaos: With Applications To Physics, Biology, Chemistry, And Engineering**

Sundaram, R. K.: **A First Course in Optimization Theory**

Sutton, D.: **Platonic Archimedean Solids**

Synge, J. L., and A. Schild: **Tensor Calculus**

Tarafdar, E. U., et al.: **Topological Methods for Set-Valued Nonlinear Analysis**

Taylor, A., and A. M. Pacelli: **Mathematics and Politics**

Taylor, J. R.: **Classical Mechanics**

Thomas Jr., G. B.: **Thomas' Calculus: Media Upgrade**

Thomas Jr., G. B.: **Calculus and Analytic Geometry**

Thompson, S. P.: **Calculus Made Easy**

Titchmarsh, E. C.: **Mathematics for the General Reader**

Todd, J.: **Basic Numerical Mathematics** (Two volumes)

Tolstov, G. P.: **Fourier Series**

Torretti, R.: **Relativity and Geometry**

Toth, G.: **Glimpses of Algebra and Geometry**

Troelstra, A. S., et al.: **Basic Proof Theory**

Vandome, A. F., et al.: **Complex Conjugate**

Vandome, A. F., et al.: **Complex Plane**

Vaught, R. L.: **Set Theory**

Velleman, D. J.: **How to Prove It**

Walker, R. C.: **Introduction to Mathematical Programming**

Wallace, D. F.: **Everything and More: A Compact History of Infinities**

Warland, D.: **Plane but Elegant Complex Numbers in Geometry**

Wasserman, R. H.: **Tensors and Manifolds: with Applications to Mechanics and Relativity**

Weatherburn, C. E.: **An Introduction to Riemannian Geometry and Tensor Calculus**

Wegener, I., and R. Pruim: **Complexity Theory: Exploring the Limits of Efficient Algorithms**

Weil, A.: **Basic Number Theory**

Wells, D. A: **Lagrangian Dynamics**

Wessel, C.: **Om directionens analytiske betegning** (Paper 1797)

Weyl, H.: **On the New Foundational Crisis of Mathematics**

Weyl, H.: **Philosophy of Mathematics and Natural Science**

White, R., and T. E. Downs: **How Computers Work**

Whitehead, A. N.: **Introduction to Mathematics**

Whittaker, E. T., and G. N. Watson: **A Course of Modern Analysis**

Wickerhauser, M. V.: **Mathematics for Multimedia**

Wilbraham, A. N. S.: **Advanced Level Complex Numbers**

Wilder, R. L.: **Introduction to the Foundations of Mathematics**

Williams, J.: **Complex Numbers (Problem Solvers)**

Winkler, P.: **Mathematical Puzzles: A Connoisseur's Collection**

Yates, R. C.: **Regular Polygons**

Yeung, R. W.: **A First Course in Information Theory**

Young, N.: **An Introduction to Hilbert Space**

Zawaira, A.: **A Primer for Mathematics Competitions**
Zill, D., and P. Shanahan: **A First Course in Complex Analysis With Applications**
Zeitz, P.: **The Art and Craft of problem Solving**

INDEX

Note: The INDEX refers to Chapter/Section and it only refers to the Chapter/Section where the entry is first mentioned.